Springer Series in

MATERIALS SCIENCE 142

AF789561

Springer Series in
MATERIALS SCIENCE

Editors: R. Hull C. Jagadish R.M. Osgood, Jr. J. Parisi Z. Wang H. Warlimont

The Springer Series in Materials Science covers the complete spectrum of materials physics, including fundamental principles, physical properties, materials theory and design. Recognizing the increasing importance of materials science in future device technologies, the book titles in this series reflect the state-of-the-art in understanding and controlling the structure and properties of all important classes of materials.

Please view available titles in *Springer Series in Materials Science* on series homepage http://www.springer.com/series/856

Lars Rebohle
Wolfgang Skorupa

Rare-Earth Implanted MOS Devices for Silicon Photonics

Microstructural, Electrical and Optoelectronic Properties

With 121 Figures

Springer

Dr. Lars Rebohle
Dr. Wolfgang Skorupa
Forschungszentrum Dresden-Rossendorf e.V.
Institut für Ionenstrahlphysik und Materialforschung
Bautzner Landstraße 400, 01328 Dresden, Germany
E-mail: l.rebohle@fzd.de, w.skorupa@fzd.de

Series Editors:

Professor Robert Hull
University of Virginia
Dept. of Materials Science and Engineering
Thornton Hall
Charlottesville, VA 22903-2442, USA

Professor Chennupati Jagadish
Australian National University
Research School of Physics and Engineering
J4-22, Carver Building
Canberra ACT 0200, Australia

Professor R. M. Osgood, Jr.
Microelectronics Science Laboratory
Department of Electrical Engineering
Columbia University
Seeley W. Mudd Building
New York, NY 10027, USA

Professor Jürgen Parisi
Universität Oldenburg, Fachbereich Physik
Abt. Energie- und Halbleiterforschung
Carl-von-Ossietzky-Straße 9–11
26129 Oldenburg, Germany

Dr. Zhiming Wang
University of Arkansas
Department of Physics
835 W. Dicknson St.
Fayetteville, AR 72701, USA

Professor Hans Warlimont
DSL Dresden Material-Innovation GmbH
Pirnaer Landstr. 176
01257 Dresden, Germany

Springer Series in Materials Science ISSN 0933-033X
ISBN 978-3-642-26558-7 ISBN 978-3-642-14447-9 (eBook)
DOI 10.1007/978-3-642-14447-9
Springer Heidelberg Dordrecht London New York

© Springer-Verlag Berlin Heidelberg 2010

Softcover reprint of the hardcover 1st edition 2010

This work is subject to copyright. All rights are reserved, whether the whole or part of the material is concerned, specifically the rights of translation, reprinting, reuse of illustrations, recitation, broadcasting, reproduction on microfilm or in any other way, and storage in data banks. Duplication of this publication or parts thereof is permitted only under the provisions of the German Copyright Law of September 9, 1965, in its current version, and permission for use must always be obtained from Springer. Violations are liable to prosecution under the German Copyright Law.
The use of general descriptive names, registered names, trademarks, etc. in this publication does not imply, even in the absence of a specific statement, that such names are exempt from the relevant protective laws and regulations and therefore free for general use.

Cover design: eStudio Calamar Steinen

Printed on acid-free paper

Springer is part of Springer Science+Business Media (www.springer.com)

To our families

Preface

Wo Licht ist, ist auch Schatten!
(*"More light, more shadow!" or simpler: "Nothing is perfect"*)
—Johann Wolfgang von Goethe, from Götz von Berlichingen, Act I

There exist already about ten books (e.g. [1–8]) – not counting the many conference proceedings volumes – on the different aspects of Si-based photonics including also the issue of silicon-based light emission. Why is now another one needed about this subject, and even more, exclusively about a special type of light emitters?

This book summarizes all aspects of the development of rare earth (RE) containing MOS devices fabricated by ion implantation as the key technology and critically reflects the related references throughout the different chapters. This work was mainly done in the course of the last 10 years. Preliminary work for this goal, undertaken mostly in the nineties, was based on the introduction of group IV elements (Si, Ge, Sn) into the thermally grown silicon dioxide leading to the highest power efficiency values in the blue–violet wavelength range. This success inspired us to use the REs as means of exploring other wavelength ranges with the same or even higher power efficiencies.

After an historical introduction of the REs and silicon-based light emission, Chap. 1 presents a review of electroluminescence from MOS-type light emitters, based on silicon and its technology. The achievement of an optimized material for electrically driven light emission, that is, efficient emission with reasonable reliability, is only possible with a deep knowledge of the materials properties determining the electro-optical (or photonic) properties (see Chap. 2). Especially intrinsic defects, hydrogen as well as the behaviour of the REs themselves were of interest. In this respect, we concentrated on the formation and the properties of clusters as a process neutralizing the emission abilities of the single RE atoms. The electrical properties of the RE-containing thermally grown silicon dioxide layers are in the focus of Chap. 3. Especially, the charge transport and trapping of electrons and holes are of special interest because it is only on the basis of knowing these properties that a deeper understanding of the reliability issues is possible. The core of the interest in the subject of this book focuses on the electroluminescence properties, which have never been worked out and reported before with the intensity presented here (see Chaps. 4 and 5). The combination of experience in MOS technology, ion implantation and RE physics and chemistry has created a unique bunch of results in the course of our work. The ion beam processing experience of the people at the research centre at Rossendorf, and partly, also at Jülich has offered the possibility to study the properties of the wide range of REs described here. Besides the EL spectra reported in Chap. 4, one of the main challenges during this work was the study and optimization of the EL efficiency as well as the exploration of different sensitization effects (see Chap. 5). Several known and newly developed methods were employed to increase the efficiency of a device concept that normally leads to strong degradation effects in the operation mode needed here, that is, hot electron generation and transport inside thermally grown silicon dioxide. The final solution of all these efforts was the introduction of the LOCOS processing in combination with the use of a dielectric double layer with silicon oxynitride on top of silicon dioxide. The related electrical degradation and EL quenching issues are described in Chap. 6.

Finally one question remains, at least from the viewpoint of an engineering driven physicist with responsibility for the society from which he gets the money for his research work: what is all this good for? First, it has to be claimed that the principal driving force for these efforts is the creation of a Si-based light source. This dates back about 20 years to the finding of Canham et al. [9] that porous silicon can emit light if excited by UV irradiation. The Si-based light source is the main obstacle to the dream of one day replacing electron/hole driven information processing by a photon driven one. This dream has pushed many people worldwide exactly these 20 years, because most of the other components necessary to realize silicon photonics, such as receivers, modulators, waveguides, etc., are available. And in the present days, 20 years is quite a long time to wait for such a development.

Looking back at the history of key technology development steps, after the invention of the wheel and the wagon more than 5,000 years ago at the dawn of the Bronze Age at several places on the globe, the next logical step to invent a self-moving wagon took several millennia until the construction of steam engine-driven locomotives at the end of the eighteenth century. The nineteenth century finished after this steam engineering period with revolutionary developments in the mastering of electricity and telegraphy/telephony for energy and information management. The field effect as the main principle of the MOS transistor dominating the Si technology up to now was invented in the 1920s by Lilienfeld [10], leading to the invention of the integrated circuit by Kilby and Noyce [11] at the end of the 1950s, that is, only about 30 years of time went by between these two important steps of technology.[1] Thus, all these point to an exponential invention cycle. After that time, a tremendous development of the device integration occurred, quite well described by Moore's law, which now leads to questions such as: "More Moore" or "More than Moore"?!

If looking from the point-contacted Ge-block of the bipolar transistor invented in 1947 at Bell Labs to the FinFet (ca. 2000) with geometrical measures in the

[1] Interestingly, J. Kilby claimed in his Nobel Price Lecture in 2000 that the idea itself, i.e. to build all electronics into one single block, was developed by G. Dummer from the English Royal Radar and Signals Establishment in the early 1950s.

low nanometre-range, one can only speak of a shocking development. It was clear that this revolution of information processing in close combination with the related materials technology led to the revolution in nano and biotechnology during the last 20 years.

But now we also have been already waiting for 20 years for the full photonic circuit with the integrated, electrically driven Si-based light source, what seems to brake the feeling about an exponential development quite a bit. The older and younger history teaches us that exponential developments sooner or later will descend into saturation or will come to a sudden end. So, are these 20 long years already an indication of saturation?

Nevertheless, our approach to use the RE-based light emitters with their limitations regarding efficiency and wear-out behaviour was devoted to applications where the light sources have to deliver certain efficiency at limited lifetime and should be of some use for the human society. This is finally described in Chap. 7 as Si-based optical biosensing using fluorescence. The check of important liquids such as drinking water or milk for estrogens and other critical substances is one of those tasks that save the human genes from too much further disturbances.

Acknowledgements

This book would not have been possible without the help of many colleagues who supported us in manifold ways ranging from experimental assistance to stimulating discussions about physics and beyond. We are especially grateful to C. Cherkouk, T. Gebel, D. Grambole, M. Helm, A. Kanjilal, J. Lehmann, A. Mrotzek, A. Mücklich, S. Prucnal, J. Sun, M. Voelskow and R. A. Yankov (presently or formerly at the Research Centre Dresden-Rossendorf); A. Nazarov, I. N. Osiyuk, I. P. Tjagulskii and S. Tjagulskii (Nat. Academy of Science, Kyiv); C. Buchal and S. Mantl (Research Centre Jülich); H. Fröb and K. Leo (Technical University of Dresden); J. Biskupek and U. Kaiser (University of Ulm); R. Frank and G. Gauglitz (University of Tübingen); B. Garrido and P. Pellegrino (University of Barcelona); T. Gregorkiewicz (University of Amsterdam) and A. Revesz (Revesz Associates, Bethesda, USA).

Dresden, Germany
August 2010

Lars Rebohle
Wolfgang Skorupa

Contents

Acronyms

AC	Alternating current
B/G ratio	Blue to green ratio
BD	Breakdown
CB	Conduction band
CCD	Charge-coupled device
CV	Capacitance–voltage
DC	Direct current
EL	Electroluminescence
ELI	Normalized electroluminescence intensity
EQE	External quantum efficiency
FA	Furnace annealing
FLA	Flash lamp annealing
FN	Fowler–Nordheim
FTIR	Fourier transform infrared
FWHM	Full width at half maximum
GeODC	Ge-related oxygen deficiency centre
HRTEM	High resolution transmission electron microscopy
IQE	Internal quantum efficiency
IR	Infrared
ITO	Indium tin oxide
IV	Current–voltage
LC	Luminescence centre
LOCOS	Local oxidation of silicon
LPCVD	Low pressure chemical vapour deposition
MOS	Metal-oxide-semiconductor
MOSLED	MOS-based light emitting device
NBOHC	Non-bridging oxygen hole centre
NC	Nanocluster
NOV	Neutral oxygen vacancy
NRA	Nuclear reaction analysis
ODC	Oxygen deficiency centre
PECVD	Plasma-enhanced chemical vapour deposition
PL	Photoluminescence

RBS	Rutherford backscattering
RE	Rare earth
RTA	Rapid thermal annealing
SOI	Silicon on insulator
TEM	Transmission electron microscopy
UV	Ultraviolet
XRD	X-ray diffraction
ESR	Electron spin resonance
QBD	charge-to-breakdown
SRIM	Stopping and Range of Ions in Matter
XPS	X-ray photoelectron spectroscopy

Symbols

A, A_{Det}	Device and detector area (Sect. 5.1)
c	Speed of light in vacuum (Sect. 4.1.2)
c_{Er}	Atomic concentration of Er (Sect. 5.3.2)
c_{O}, c_{N}	Atomic concentration of oxygen or nitrogen (Sect. 2.2.2)
d	Si or Ge NC diameter (Chap. 5)
d_{eff}	Effective thickness of a dielectric stack (Sect. 3.1.3)
d_{N}, d_{Ox}	SiON and SiO_2 layer thickness (Sect. 3.1.3)
D	Dielectric displacement (Sect. 3.1.3)
D	Implantation dose (Sect. 5.1.1)
E	Electric field
E_{G}, E_{G0}	Bandgap energy of a Si NC and bulk Si (Sect. 5.3.1)
E_{Ox}	Electric field in the SiO_2 layer (Sect. 3.1.3)
E_{Ph}	Photon energy
f	Fraction of excited LCs
F	Cumulative failure rate (Sect. 6.1.2)
h	Planck's constant
$\hbar$	Reduced Planck's constant $\hbar = h/2\pi$ (Sect. 3.1.1)
j	Current density
J	Total angular momentum (Sect. 4.1.1)
k_i	Transition rate (Chap. 5)
L	Total orbital angular momentum (Sect. 4.1.1)
m^*	Effective electron mass (Sect. 3.1.1)
m_0	Electron rest mass (Sect. 3.1.1)
N^*	Total number of *excited* RE^{3+} ions (Sect. 4.2.1)
N_{BD}	Critical defect density to induce an electrical breakdown (Sect. 6.2.1)
N_i	Number of RE^{3+} ions or of sensitizers being in a certain state (Sect. 5.1.3)
N_{max}	Total number of *excitable* RE^{3+} ions (Sect. 4.2.1)
N_{LC}, N_{LC}^0	Current and initial number of active LCs (Sect. 6.2.1)
N_{RE}	Total number of RE^{3+} ions coupled to a sensitizer (Sect. 5.1.3)
N_{S}	Total number of sensitizers coupled sensitizers (Sect. 5.1.3)
P	Trapping efficiency (Sect. 3.1.3)

P_{Det}	Optical power measured by the detector (Sect. 5.1)
P_{G}	Defect generation probability (Sect. 6.1.1)
P_{opt}	Total optical power emitted by a light emitter (Sect. 5.1)
q	Elementary charge (Sect. 3.1)
Q_{BD}	Charge-to-breakdown (Chap. 6)
Q_{IF}	Trapped charge at the Si–SiO_2 interface (Sect. 3.1.3)
Q_{inj}	Injected charge
Q_{N}, Q_{Ox}, Q_{ON}	Trapped charge in the SiON layer, the SiO_2 layer, and at the interface of both
Q_{t}, $Q_{\text{t}}^{\text{max}}$	Trapped and maximum trapped charge (Sect. 3.1.3)
r	Distance between device and detector (Sect. 5.1.2)
R_{Ox}	Relative oxygen ratio (Sect. 2.2.2)
S	Total spin (Sect. 4.1.1)
t	Time
t_{m}	Modal lifetime of a device (Sect. 6.1.2)
T_{a}	Annealing temperature (Sect. 5.3.2)
V_0	Applied gate voltage
$\bar{x}_{\text{N}}, \bar{x}_{\text{Ox}}$	Charge centroid in the SiON and the SiO_2 layer (Sect. 3.1.3)
α, α^*	Fowler–Nordheim constants defined by (3.1) and (3.2)
α, α^*	Cumulative transition rate of a sensitizer (Sect. 5.1.3)
β	Observation angle (Sect. 5.1.2)
β	Cumulative transition rate of a RE^{3+} ion (Sect. 5.1.3)
ΔV_{CC}	Change of the applied voltage under constant current injection
ΔV_{FB}, ΔV_{MG}	Change of the flatband and midgap voltage (Chap. 3)
$\Delta V_{\text{FB}}^{0}, \Delta V_{\text{FB}}^{\text{RE}}$	Change of the flatband voltage for an unimplanted and a RE implanted device (Sect. 3.2)
ΔV_{IV}	Voltage shift of the *IV* characteristic (Chap. 3)
$\Delta V_{\text{IV}}^{0}, \Delta V_{\text{IV}}^{\text{RE}}$	Voltage shift of the *IV* characteristic of an unimplanted and a RE implanted device (Sect. 3.2)
ε_0	Dielectric constant
ε_{N}, ε_{Ox}	Relative dielectric constant of SiON and SiO_2 (Sect. 3.1.3)
η_{EQE}, η_{IQE}	External and internal quantum efficiency
η_{PE}	Power efficiency
θ	Ratio between the maximum EL intensity of an Eu-implanted MOSLED achieved during its operation lifetime and the initial EL intensity (Sect. 6.2.5)
θ_{B}	Bragg angle (Sect. 5.3.2)
λ	Wavelength
ρ	Trapped charge distribution (Sect. 3.1.3)
σ_{D}	Dispersion parameter for the Weibull distribution (Sect. 6.1.2)
σ_{EL}	EL excitation cross section (Chap. 4)
σ_i	Trapping cross section
σ_{Q}	EL quenching cross section (Sect. 6.2.1)
τ	EL decay time
τ_{Q}	Characteristic EL quenching time (Sect. 6.2.1)
ϕ_0	Barrier height for electron injection (Sect. 3.1.1)

Chapter 1
Silicon-Based Light Emission

1.1 Historical Remarks

There are at least three different fields that have contributed to the development of RE-implanted MOS devices, namely the RE elements and their unique optical properties (see also Sect. 5.1), the development of the MOS technology and the field of Si-based light emission.

RE elements have aroused interest for more than 200 years. Their discovery is closely linked with two locations in Sweden, namely the small town of Ytterby and the ore field of Bastnäs, both being the origin of the minerals from which most of the RE elements were extracted in the nineteenth century. The story started a little earlier, in 1787, when Carl Axel Arrhenius found a black mineral, hitherto unknown in the region surrounding Ytterby. On examining this mineral, the Finnish chemist Johan Gadolin found yttria, the oxide of the new element yttrium,[1] in 1794. Shortly after, in 1803, another oxide, ceria, was discovered independently from each other by Berzelius and Hisinger in Sweden and Klaproth in Berlin in a mineral originating from Bastnäs. After a break of more than 30 years, between 1839 and 1843, Karl Gustav Mosander expanded the palette of known RE elements by lanthanum, terbium, erbium and didymium. However, separating chemical elements such as the REs, which are so similar, was extremely difficult at that time, and it is no wonder that it took several decades to discover that didymium was, in fact, a mixture of the elements praseodymium and neodymium.

With the introduction of spectral analysis and the periodic system of elements, the situation became better in the middle of the nineteenth century, and more RE elements were discovered: ytterbium (Marignac 1878), scandium (Nielsen 1878), samarium (Lecoq de Boisbaudran 1879), holmium and thulium (Cleve 1878/79), and gadolinium (Marignac 1880). However, the history of these discoveries is cloudy, and it is sometimes difficult to identify the 'true' discoverer. For instance, holmium was simultaneously discovered by Delafontaine and Soret, and first indications of the existence of samarium date back to 1853 when Marignac investigated

[1] By definition, the REs consist of the elements scandium, yttrium, and the 15 lanthanides.

the absorption lines of what was called didymium. In the following decade, the discovery of new RE elements – with the exception of promethium and lutetium – was completed with the detection of praseodymium, neodymium (both Auer 1885) and europium (Lecoq de Boisbaudran 1890).

This small excursion into history can be referred to in much more detail in [12]. The optical properties of the REs were utilized early on. One of the first commercially successful optical applications was the use of cerium oxide as an additive in the gas mantle of a so-called Auer lamp in 1891 [12]. Today, RE elements are involved in countless optical applications: they are part of many dyes and phosphors, they have strongly influenced the development of laser technology and they serve as luminescence centres (LCs) in numerous EL devices. Not to forget their important role in 'electromagnetic' applications as rare earth magnets and superconductors, and their significance as trace elements in geochemistry.

As is well known, the invention of the bipolar point-contact transistor in 1947 by John Bardeen and Walter Brattain, and that of the bipolar junction transistor by William Shockley one year later heralded the information age with the domination of the Si technology. As there are many excellent publications reflecting this history [13, 14], it is not necessary to add another view on it. The only event we want to mention is the invention of the Si MOSFET by Atalla and Kahng [15], in 1959. In the following two decades, the material system Si–SiO_2–Al was investigated in great detail, and it belongs to the best-known material systems today.

Si-based light emission became popular with the discovery of orange PL in porous Si in 1990 [9], although there were a couple of publications that had reported EL from Si-based materials earlier [16]. This included the observation of EL from SiC as early as 1907 [17] and the 1920s [18], from the Si and Ge bandgap [19] and from electron–hole recombination in Si p–n junctions [20]. These early works did not make much impact in the scientific community, but in the 1990s, the limitations of pure electronic data processing (speed limitations, cross-talk, etc.) became more and more apparent. This generated a strong interest in photonics, and especially in Si-based photonics, with the vision to replace parts of the electronic data processing by optical counterparts. Among the different components of a photonic circuitry, the Si-based light emitter is one of the key elements. However, as discussed in Chap. 7, its range of potential applications goes far beyond the field of optical data processing.

In the last two decades, a couple of Si-based materials have become interesting with respect to light emission, which was usually observed in initial PL investigations. Thus, PL from Er in Si had been found quite early [21, 22]. After 1990, the field of Si-based light emission quickly expanded to other material systems including Er in SiO_2 [23], Si-rich [24–26] or Ge-rich SiO_2 layers [27–29] and $FeSi_2$ precipitates in Si [30, 31]. Subsequent investigations were engaged in dislocations in Si [32] (although this topic is much older), Si/SiO_2 superlattices [33], Si/Ge superlattices [34, 35] and in the incorporation of RE elements other than Er in SiO_2 layers [36]. Also, the 'ancient' topic of EL from p–n junctions in Si experienced a renaissance [37–39].

At the turn of the millennium, the detection of gain in Si-rich SiO_2 [40] stimulated the idea of a Si-based laser, and today, the Raman silicon laser [41, 42] is probably very close to this vision. Another, quite recent approach is the use of Ge on Si for wavelengths of about 1,600 nm [43]. More details about the current status of Si laser development can be found in [44]. More recent developments also consider the use of sensitizers to enhance the luminescence efficiency of Er and other RE elements, which is discussed in Chap. 5.

1.2 Electroluminescence from Si-Based, MOS-Type Light Emitters

Although strong photoluminescence was observed in a number of studies, it was difficult to excite these structures electrically, and the number of studies that report efficient EL are significantly lower than those dealing with photoluminescence only. However, even this field is large, for which reason this short chapter on Si-based EL is limited to MOS-type light emitters that use SiO_2 or modified SiO_2 on Si.

The first studies of EL properties of MOS systems with stoichiometric or Si-rich SiO_2 served as an additional tool for the investigation of their physical properties and their electrical quality [45]. In case of stoichiometric SiO_2, the EL is either caused by intrinsic defects excited by the impact of hot electrons or by electron–hole recombination in Si. The latter is usually the case for very thin oxide layers, which often deliver a weak EL signal of around 1,100 nm due to electron–hole bandgap recombination in Si [46–52]. Interestingly, this EL can be enhanced by using a suspension of SiO_2 nanoparticles, instead of a closed SiO_2 layer [53]. However, both visible EL due to hot electrons crossing the SiO_2 layer [48] and sub-bandgap EL at 1.5 μm originating from dislocation loops [54] were also observed. In addition, visible EL from blue to red, depending on the applied RTA processing, was found in native SiO_2 sandwiched between a p-type Si substrate and an Au electrode [55].

The EL of Si-rich SiO_2 is usually much stronger than that of stoichiometric SiO_2 and can be classified either according to the spectral region where the EL appears or according to the type of LC, namely point defects related to Si excess, Si NCs or defects at the NC surface. Red or NIR EL found in Si-rich SiO_2 is mostly attributed to exciton recombination in Si NCs [56–65], whereas green or blue EL is generally ascribed to defects [66–72]. However, red EL is sometimes also attributed to defects [73], especially if the peak positions do not vary with the fabrication conditions. Quite often, only a red or NIR EL is observed when a poly-Si gate is used because of the fairly high absorption coefficient of poly-Si for shorter wavelengths. In contrast to this, blue or green EL lines, either alone or together [74–77] with the red or NIR EL, are found when ITO or thin, semitransparent metal films are used. Of course, the question as to which type of EL is dominant, strongly depends on the Si excess concentration, the annealing and other fabrication details. Some publications [78–80] reported that the application of an AC or pulsed voltage can be more advantageous than the DC operation mode. This especially applies to the case of exciton recombination in Si NCs as the AC mode, depending on its frequency, allows the subsequent injection of charge carriers of opposite polarity within a short time scale. On comparing different types of Si NCs, it was found that small NCs are more efficient than larger ones [60], and that amorphous NCs require a lower voltage but a higher injection current to achieve the same EL intensity as crystalline NCs [81]. A model of the charge transport between NCs and a simulation of the resulting EL properties is given in [82], and a small review about EL in SiO_x films can be found in [83]. A more detailed discussion about the typical defects in ion-implanted SiO_2 and the EL lines they may account for can be found in Sects. 3.2.1 and 5.1.3.

Another specific version of Si-rich SiO_2 are Si/SiO_2 superlattices, where the Si layers are either still closed layers or layers decomposed in Si NC because of high temperature annealing. EL from such structures, particularly in the red spectral region, was observed [84–94] and explained by hot electron relaxation [87] or recombination of injected electrons and holes within Si NCs and/or via LCs near Si/SiO_2 interfaces in the superlattice [92]. More details can be found in the review of Zheng et al. [95].

Instead of Si, other group IV elements can be incorporated into the SiO_2 layer. For carbon- and Si-rich SiO_2, a broad blue-green EL was reported [63, 96, 97] and assigned to $Si_yC_{1-y}O_x$ complexes or carbon-rich nanoprecipitates. Together with the red/NIR EL from Si NCs, the EL may appear to be white [63, 97]. The appearance of a UV peak at 3.2 eV (387 nm) was related to the formation of silicon-carbide clusters [98]. Strong blue-violet EL of around 400 nm was observed from Ge-implanted SiO_2 layers [66, 67, 99–103], which was significantly higher than that from Si- or Sn-implanted SiO_2 [67]. Thereby, the blue-violet EL is caused by a radiative triplet–singlet transition within a Ge-related NOV. However, green, red and NIR EL from Ge-rich SiO_2 layers are also known [101, 104–110]. Another interesting contribution is the observation of red EL from a Ge/SiO_2 superlattice within an MOS structure [112].

The literature on EL from RE-implanted MOS structures can be roughly divided into two groups: those that incorporate Er and those that also use other RE elements. EL of around 1.5 μm from Er-doped SiO_2 [36] and Er-doped Si-rich SiO_2 [113] was reported in 1997, followed by a number of other publications, including [114–122]. Since 1997, EL has also been reported for Tb-implanted SiO_2 layers [123–125], and in 2005, an EL power efficiency of 0.3% was achieved for such layers [126]. In 2004 strong EL from Gd-implanted SiO_2 was found [127], and in 2007 it was shown that the EL of Eu-implanted SiO_2 can change its colour from red to blue and vice versa, depending on the electrical excitation conditions [128]. There are further reports about the EL from Ce- [129, 130], Eu- [131, 132], Gd- [133–140], Tb- [126, 141, 142], Tm- [129, 143, 144] and Yb-implanted SiO_2 layers [145–147].

Chapter 2
Microstructure

The Si-based light emitters to be discussed in this book are based on an RE-implanted, standard MOS structure with an additional dielectric layer, also known from MNOS (metal–nitride–oxide–semiconductor) devices. There is a giant pool of literature about Si technology, Si processing, the structure and point defects in SiO_2 and related materials, as well as about the structural properties of the Si–SiO_2 interface. Consequently, Sect. 2.1 reports only on the specific fabrication and device parameters used for the light emitter preparation. Section 2.2 touches briefly on the topic of point defects in SiO_2 and SiON as well as the issue of RE diffusion under annealing, while Sect. 2.3 concentrates on the formation and the evolution of clusters.

2.1 Device Structure and Fabrication

2.1.1 Layout

Usually, the Si-based light emitters are prepared by standard silicon MOS technology and consist of a layer stack with Si substrate at the bottom, followed by a thermally grown gate oxide, a SiON layer deposited by PECVD and a transparent front electrode made of ITO. Although the gate oxide is the active medium that contains the RE LCs, the SiON layer is a dielectric protection layer, which strongly enhances the operation lifetime of the devices [142, 148]. The latter point is comprehensively discussed in Sect. 6.3. Generally, a gate oxide thickness of 100 nm is used, although for some specific investigations, a 200-nm thick SiO_2 layer or thinner layers with a thickness ranging between 10 and 75 nm were used. Typical SiON layer thicknesses range from 50 to 200 nm but amount to 100 nm in most cases. If the light emitter is intended to be used in humid or wet media, the device can be passivated by a thick SiO_2 layer deposited by PECVD, which leaves only the contact pads open. The rear electrode made of aluminium is fabricated on the backside of the Si wafer. In the following sections, such a structure is called MOS light emitting device (MOSLED).

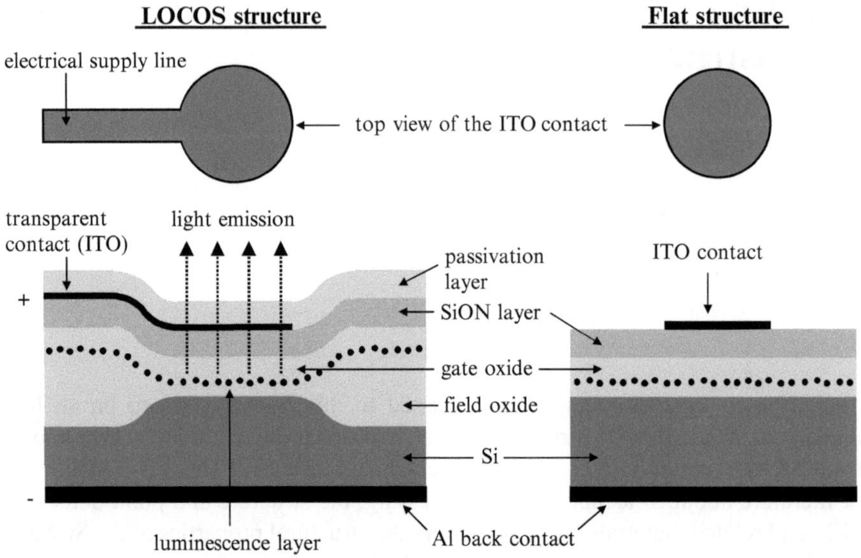

Fig. 2.1 Basic designs of RE-implanted MOSLEDs in the form of a LOCOS structure (*left*) or a flat structure (*right*) together with a *top view of the upper supply line*

As shown in Fig. 2.1, the light emitters were fabricated using two different designs. The more complex structure is processed in local oxidation of silicon (LOCOS) technology, (see [149, 150] for more details) with the gate oxide being surrounded by a thicker field oxide. This design allows the processing of electrical supply lines together with the upper ITO electrode as applying a voltage between the upper and lower electrodes would result in a larger electric field across the thin gate oxide than across the thick field oxide. As a result, only the gate oxide area emits light and not the area of the electrical supply line. In addition, the bird beak-shaped transition between the field and gate oxide avoids an early electrical breakdown at the edge of the electrode and thus prolongs, together with the SiON layer, the operation lifetime of the devices. Because of these advantages, the LOCOS design of the light emitters is usually applied in real applications. However, the LOCOS design is troublesome for a couple of electrical and microstructural investigations. If the capacitance of the MOS structure has to be measured, the signal is always composed of the component of the light emitting area and the contribution of the electrical supply line. In case of microstructural investigations such as TEM, RBS and AES, advanced preparation techniques (e.g. focused ion beam preparation) are required to hit the light emitter area precisely. Sometimes even the ITO electrode is perturbing, e.g. for RBS because of mass interferences between indium and the RE elements. For such investigations, a flat structure is used, as schematically drawn on the right of Fig. 2.1.

The fabrication of the light emitters starts with the LOCOS process, including the thermal growth of the gate oxide by dry oxidation at 1,050°C, followed by ion implantation of RE elements and a thermal treatment. The details of the implantation and the thermal treatment are discussed in the subsequent chapters. The processing is continued with the deposition of a SiON layer by PECVD at a temperature of 325°C, followed by exactly the same thermal treatment performed after the ion implantation. All the implications connected with implantation-induced defects in SiO_2 and the hydrogen load of SiON are discussed in Sects. 2.2.1 and 2.2.2, respectively. After the second thermal treatment, the Al back electrode is sputtered in combination with a contact annealing at 400°C, followed by magnetron sputtering of ITO on the front side of the wafer. It is worth noting that the use of Si_3N_4 instead of SiON may cause adhesion problems of the ITO layer. The upper electrode is structured by means of lithographic patterning, whereas the size, the shape and the arrangement of the light emitters are only limited by the possibilities of photolithography. If not otherwise stated, the electrical and optical results presented in this work were obtained from circular dots with a diameter of 300 μm, although feature sizes down to 2 μm were tested. Figure 2.2 gives a brief impression of what is technically feasible.

2.1.2 Rare Earth Implantation

Ion implantation is an essential processing step that determines the special type of light emitter by the RE element and influences the operation lifetime and efficiency by the implantation dose. The RE ions, namely the elements Ce, Pr, Eu, Gd, Tb, Er

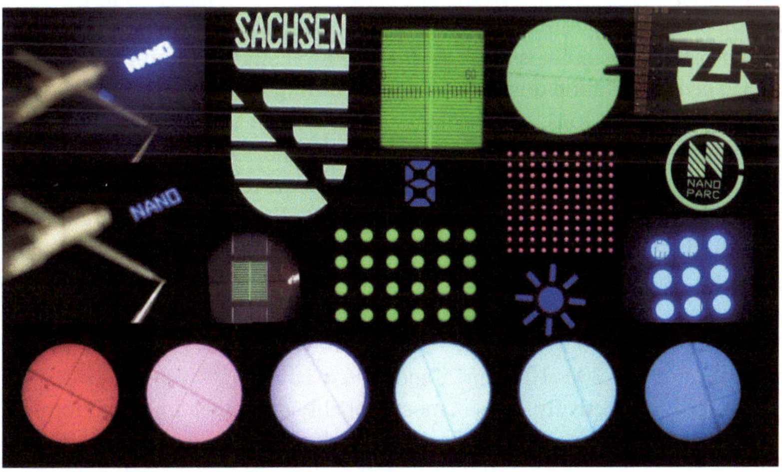

Fig. 2.2 Light emitting Ge- and RE-implanted MOS structures with different layouts. The *blue*, *green* and *reddish* structures in the *upper part of the figure* originate from Ge-, Tb- and Gd-implanted MOSLEDs. The *lower row* displays Eu-implanted MOSLEDs, with different Eu concentrations processed under different annealing conditions

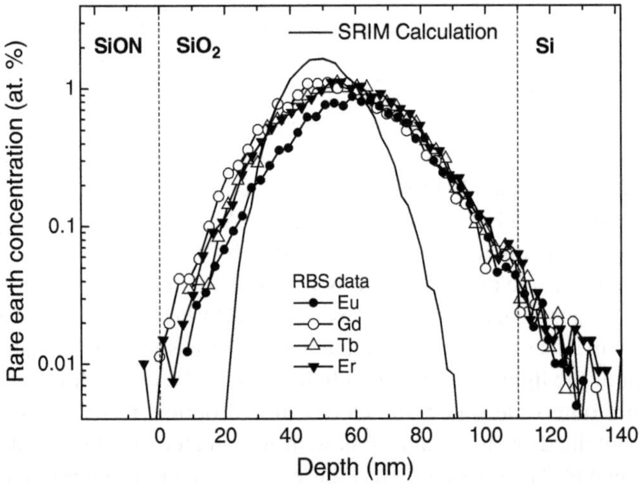

Fig. 2.3 Elemental depth profiles calculated for 100 keV Tb by SRIM (*solid line*) and profiles calculated from RBS spectra for Eu, Gd, Tb and Er in SiO_2 implanted with a nominal dose of $3 \times 10^{15}\,cm^{-2}$. The nominal peak concentration of Tb was 1.5%

and Yb, were single implanted in such a way that the projected range of implantation R_P is approximately in the middle of the gate oxide layer, which corresponds to an implantation energy of about 100 keV for an oxide thickness of 100 nm. Some attention should be paid to the fact that not too many ions are implanted into the Si–SiO_2 interface so as to keep its excellent electrical quality. In addition, RE ions that are located too close to the injecting interface will not take part in the electrical excitation process as outlined in Sect. 5.2.2. According to SRIM calculations [151], the depth profile has a Gaussian-like shape with a slight asymmetry as shown in Fig. 2.3. The implantation dose was varied according to the desired peak concentration of the implanted ions, which ranged from 0.25 to 9%. Optimum results were mostly achieved for RE concentrations between 1 and 3%.

However, there are already slight differences in the as-implanted profile between the prediction of SRIM and the profiles calculated from RBS spectra. As shown in Fig. 2.3, the RBS profiles are significantly broader, and this is accompanied by an adequate reduction of the peak concentration. Furthermore, the total dose of implanted ions as calculated from RBS spectra deviates in the order of 10% from the nominal value of $3 \times 10^{15}\,cm^{-2}$. In detail, total doses of $2.65 \times 10^{15}\,cm^{-2}$, $3.44 \times 10^{15}\,cm^{-2}$, $3.33 \times 10^{15}\,cm^{-2}$ and $3.46 \times 10^{15}\,cm^{-2}$ were measured for Eu, Gd, Tb and Er, respectively. This is most probably due to the implantation-induced diffusion processes that may occur under high implantation currents. Nevertheless, the value of the nominal peak concentration is used in the following to avoid confusion.

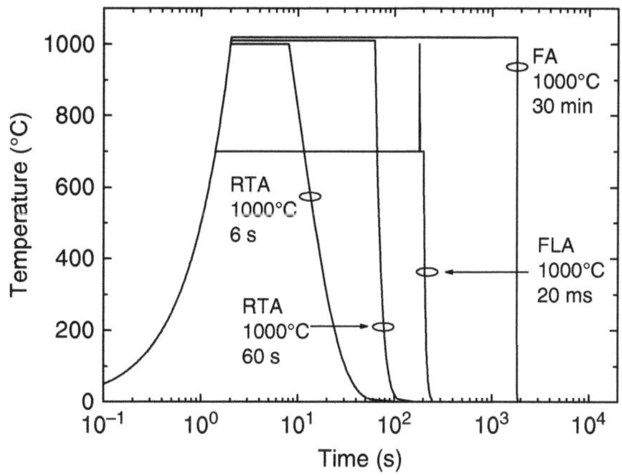

Fig. 2.4 Schematic temperature profiles of different types of annealing. The plateau at 1,000°C is slightly split for better projection

2.1.3 Annealing Conditions

Appropriate choice of annealing conditions strongly determines the operation lifetime and efficiency of the potential light emitter. On the one side, annealing is necessary to remove implantation-induced defects, to anneal defects introduced by the PECVD deposition and to activate the RE LCs. On the other, large-scale redistributions such as diffusion or the formation of huge clusters have to be avoided. The RE-implanted MOSLEDs were subjected to several types of annealing, namely furnace annealing (FA), rapid thermal annealing (RTA) and flash lamp annealing (FLA), which cover a wide range of annealing times. Figure 2.4 shows the schematic temperature profiles of the different types of annealing with an arbitrary ramp and cooling phase in order to illustrate the different physical meanings of the annealing times. For the RE-implanted MOSLEDs discussed in this book, mainly five types of annealing were used, but with different annealing temperatures: FA for 30 min, RTA for 6 or 60 s and FLA for 0.6 or 20 ms. Although for FA 30 min and RTA 60 s, the ramp and cooling phase is negligible, for RTA 6 s it forms a significant part of the temperature profile. The temperature load is therefore higher than in the ideal case of a rectangular profile. This point is even more critical in the case of FLA. First, a preheating, lasting several minutes (at 700°C for 3 min in our case), is usually applied before the flash is added. Second, the annealing time of 0.6 and 20 ms denotes the width of the applied flash pulse, but the cooling occurs, obviously, on a much longer time scale. Nevertheless, the thermal budget applied to a sample declines undoubtedly with decreasing annealing time, but possibly not exactly by the factor the different annealing times imply. Further information about RTA and FLA technology can be found in [152, 153]. As already mentioned in Sect. 2.1.1, a specific type of annealing was always applied twice to the light emitters.

The term "thermal budget" denotes a certain combination of annealing time and annealing temperature, but is not exactly defined in the literature. Two thermal budgets with temperatures T_1 and T_2 as well as annealing times t_1 and t_2 can easily be compared if both relations $T_1 > T_2$ and $t_1 > t_2$ are valid or vice versa. However, combinations such as $T_1 > T_2$ and $t_1 < t_2$ are much more problematic. A simple, traditional definition for the thermal budget is the product of time and temperature, or more generally, the area under the curves in Fig. 2.4. However, this definition is impractical in most cases as, for example, the effect of an annealing with $T_1 = 400\,\mathrm{K}$ and $t_1 = 4$ min is hardly the same as that with $T_1 = 1{,}600\,\mathrm{K}$ and $t_1 = 1$ min. It is therefore more reasonable to define the thermal budget according to the physical effect the application of a certain thermal budget will cause. Assuming that the effect of annealing temperature and time is similar to the case of diffusion, the thermal budget might be proportional to $\sqrt{t_i}\exp(-E_a/kT_i)$, with E_a being general activation energy. If the ratio between two thermal budgets is used to decide which one is higher, the result will depend on the specific value of E_a. Because of this ambiguity, further discussion on thermal budgets is conducted qualitatively, which is, however, sufficient to outline the main tendencies of the behaviour of RE-implanted MOSLEDs. A more detailed discussion about thermal budgets for short-time annealing can be found in [154, 155].

2.2 Point Defects and Diffusion

2.2.1 *Intrinsic and Implantation Induced Defects in SiO_2*

SiO_2 as the most essential dielectric in microelectronics and its interface to Si is one of the best investigated material. There have been numerous surveys of the structure and point defects of amorphous SiO_2 [156–162]. Shortly, amorphous SiO_2 consists of tetrahedron rings with a Si atom in the centre and four oxygen atoms at the corners of the tetrahedron. The number of tetrahedrons in the ring and the angle between two tetrahedrons can range from 3 to 8 and 120° to 180°, respectively, with the most probable value of 6 ring members and 144° for the Si–O–Si bond angle of amorphous SiO_2 [156]. Because of the higher atomic concentration of Si atoms in crystalline Si compared with SiO_2, there is a transition zone extending up to 3 nm from the Si–SiO_2 interface into the SiO_2. Within this zone, tetrahedron rings with 3–4 members dominate, and the concentration of the Si^{3+} suboxide measured by XPS increases towards the interface [156].

Defects in SiO_2 can be generally classified as to whether they are oxygen deficient or excessive in nature [163]. Figure 2.5 displays the ≡Si–O–Si≡ bond in the undisturbed SiO_2 network and possible defects in the oxide matrix. If the bond between one oxygen and one Si atom of the SiO_2 network is broken, the E' centre ≡Si• and the non-bridging oxygen hole centre •O–Si≡ may arise, which are both visible with electron spin resonance measurements, due to the unpaired spin of the single electron. The E' centre, which exists in many modifications depending on its chemical environment [158], is believed to be a precursor of the NOV. Another possibility is the formation of the twofold coordinated Si-atom Si_2^0, which requires

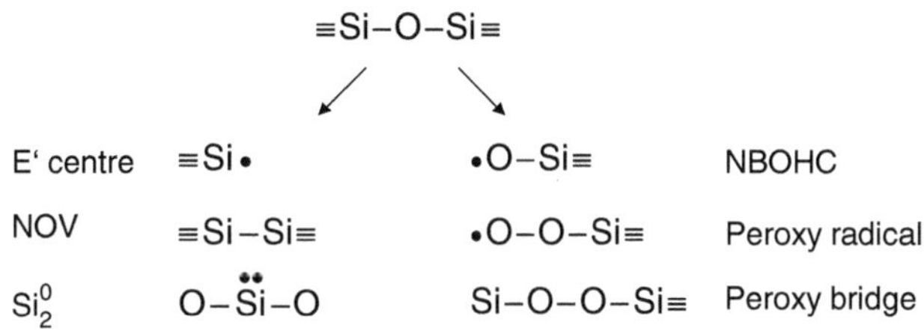

Fig. 2.5 Basic scheme of possible defects in the SiO_2 (after [67] and [163])

the breaking of two Si bonds. By binding an OH group followed by hydrogen dissociation, the non-bridging oxygen hole centre can transform into a peroxy radical or peroxy bridge. In the E' centre, the NOV and Si_2^0 are assigned to the category of ODCs, whereas the NBOHC and the peroxy structures are regarded as oxygen excess defects together with interstitial oxygen in the form of O_2 and ozone. Furthermore, there are additional types of ODCs with more than two Si atoms. Mostly, the defects can convert into each other, and the possible termination of dangling bonds by hydrogen adds another degree of freedom. Finally, small Si nanoclusters can be formed under high temperature annealing.

The SiO_2 network is heavily damaged during the RE implantation. Figure 2.6a shows the energy deposition during a 100 keV Tb implantation with a nominal dose of $3 \times 10^{15}\,cm^{-2}$, according to the energy transfer of the Tb ions to electrons, phonons and atoms of the target material. The first two processes lead mainly to bond breaking and the heating of the target, while the last process displaces the target atoms. The dashed horizontal line marks the amount of energy deposition that is required to cause maximum damage in amorphous SiO_2 [159, 165] and to amorphize crystalline Si [164]. As seen in Fig. 2.6a, the SiO_2 network is completely destroyed in the front part of the SiO_2 layer, and even the Si close to the Si–SiO_2-interface is exposed to an energy deposition, which is not enough for a complete amorphization but sufficient to cause a significant damage. During the displacement of a target atom, a significant momentum in the forward direction is also transferred to the displaced atom, leading qualitatively to a vacancy and interstitial distribution, as shown in Fig. 2.6b. Although these profiles overestimate the real concentration of vacancies and interstitials because of the partial reconstruction of the SiO_2 network during the implantation process, the qualitative picture is obvious. These profiles demonstrate that there are deviations from the stoichiometric Si:O ratio of SiO_2, implying the existence of regions with a local excess of Si or oxygen and the formation of the corresponding defects. In fact, typical defects found after implantation are the NOVs, E' centres, peroxy radicals and interstitial O_2 [166]. Additionally, this ensemble of defects is extended by defect types caused by the implanted RE ions.

With increasing annealing temperature, the different types of defects are successively annealed out. So the concentration of E' centres measured by ESR decreases with increasing annealing temperature and almost disappears at 600°C [167]. The

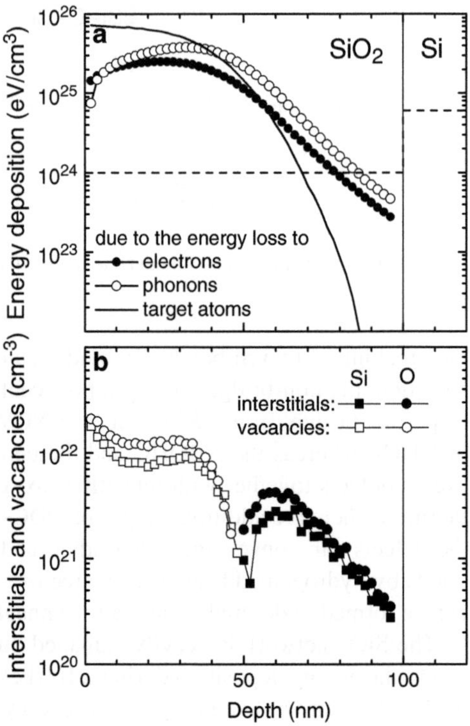

Fig. 2.6 Energy deposition (**a**) and resulting profiles of Si and O interstitials or vacancies (**b**) according to a full damage cascade calculation by SRIM [151] for a 100 keV Tb implantation into 100 nm SiO_2 on Si, with a nominal dose of 3×10^{15} cm^{-2}. The *dashed line* in (**a**) represents the value of energy deposition, which is necessary to amorphize Si [164] or to cause maximum damage in SiO_2 at room temperature [159, 165]

same temperature was needed to anneal out implantation-induced defects, as probed by positron annihilation spectroscopy [168]. However, especially for heavier ions, higher temperatures are needed, which probably have to reach the original growth temperature of the SiO_2 layer for an optimum recovery of the SiO_2 network [165]. In addition, the implanted RE ions introduce a non-stoichiometry and compete with Si for oxygen, for which reason ODCs will remain in the oxide layer even after FA 1,000°C. Furthermore, in case of RE or RE oxide cluster formation, there exists a shell of structural defects around the clusters as they do not fit perfectly into the SiO_2 network.

2.2.2 *Hydrogen and Other Defects in SiON*

The properties of SiON deposited by PECVD strongly depends on the fabrication conditions, namely on the deposition temperature, the gas flow ratio, the pressure, the HF power and even the geometry of the plasma chamber. Nevertheless, there are some general properties that are briefly discussed in the following section, together with the results of our measurements. Similar to SiO_2, the network of $Si_xN_yH_z$ is also based on Si-centred tetrahedrons, but with N or Si atoms at the corners [169].

As proved by X-ray absorption near-edge spectroscopy, the Si-centred tetrahedron with N or O atoms at the corners is the basic structural element of the network of SiON layers produced by PECVD [170]. There is a significant number of dangling Si or N bonds, which are frequently denoted in the literature as K and N centres, respectively. Because of the variability of the local network structure, there is a large family of different K centre types, as illustrated in [171]. The use of SiH_4 during deposition introduces larger amounts of hydrogen into the layers, which results in a termination of the dangling bonds by hydrogen. In case of SiON, a fraction of the N sites is occupied by oxygen. In addition, oxygen also substitutes hydrogen bonded to Si, which results in a much higher number of N–H bonds compared with the number of Si–H bonds [171]. As in the case of SiO_2, missing oxygen or nitrogen atoms can give rise to the formation of NOV defects [172, 173], whose number, however, is assumed to be low in the case of stoichiometric SiON (i.e. no Si excess).

Clearly, the hydrogen content of a SiON layer deposited by PECVD or LPCVD decreases with higher deposition temperature [174] or with a post-deposition thermal treatment at temperatures above the deposition temperature [175–177], but there is a discussion in the literature on whether this improves the quality of the SiON or not. In some cases, a compaction of the SiON layer with a shrinking of the layer thickness and the danger of crack formation is observed [178]. However, the removal of hydrogen is not accompanied by an adequate increase of N and Si dangling bonds because of the cross linking effect, which leads to the formation of new Si–N bonds, as measured by FTIR absorption spectroscopy [175]. It was reported that the density of paramagnetic defects in SiON traced by electron spin resonance generally decreases with annealing but shows a minimum in the temperature range of 500–600°C [176]. Other investigations revealed that the electrical quality of thin SiON layers improved after RTA 900°C or FA 900°C, in terms of lower interface trap densities and lower leakage currents [179].

Figure 2.7 compares the hydrogen content of unimplanted and Eu-implanted MOSLEDs for different annealing conditions. The hydrogen depth profile was determined by nuclear reaction analysis (NRA), exploiting the resonant reaction $^{15}N\,(6.385\,MeV) + {}^{1}H \rightarrow {}^{12}C + {}^{4}He + \gamma$-rays (4.43 MeV) by increasing the incident energy of ^{15}N ions and thus moving the resonance at 6.385 MeV progressively to greater depths. In all the cases, the hydrogen profiles are characterized by a nearly constant concentration in the SiON layer, which drops down at the SiO_2–SiON interface by a factor of 3–10 depending on the type of annealing. In the as-implanted state, a hydrogen concentration of about 18% is measured in the SiON layer, whereas this value decreases to 3% and 0.2% in the case of FLA 1,000°C and FA 1,000°C, respectively. The comparison between Eu-implanted devices (closed symbols) and unimplanted devices (open symbols) reveals that (1) the hydrogen content strongly decreases with an increasing thermal budget, as expected from the literature and (2) that this process is entirely independent of whether the MOSLED was implanted or not.

The loss of hydrogen is accompanied by a compaction of the SiON layer, which was investigated by ellipsometric measurements. Figure 2.8 shows the relative thickness and the refraction index of the SiON layer at a wavelength of 633 nm as a

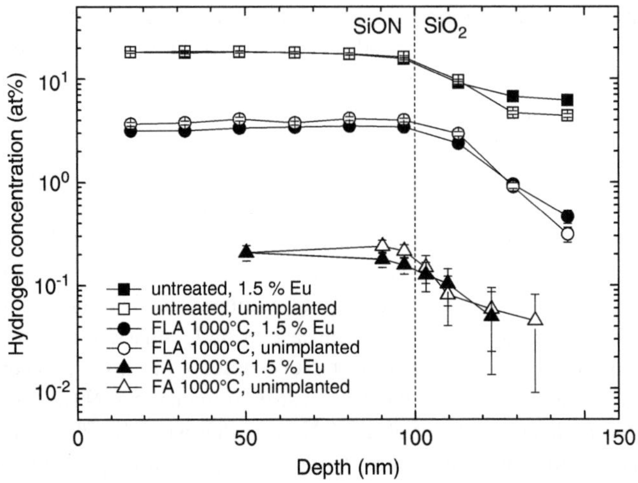

Fig. 2.7 Elemental depth profiles of hydrogen in unimplanted and Eu-implanted MOSLEDs, as derived from NRA. A SiO_2 density of $2.2\,g\,cm^{-3}$ and an average energy loss of $1.55\,keV\,nm^{-1}$ have been assumed to calculate the depth scale (after [132])

function of the annealing time at 1,000°C (a) and the annealing temperature (b). The as-deposited SiON layer shrinks by 13–14% if subjected to FLA 1,000°C 20 ms or FA 700°C, followed by a more shallow decrease of the layer thickness for higher thermal budgets. Concurrently, the refraction index increases from 1.73 to 1.76–1.78, followed by a further increase to 1.82 for higher thermal budgets. In a first approximation, the relative compaction of the SiON layer seems to be independent of the SiON layer thickness. The compaction will give a minor but significant contribution to the interpretation of electrical results, which is discussed in the corresponding chapters.

The composition of the layer system Si–SiO_2–SiON was traced by AES measurements as illustrated in Fig. 2.9. As seen, the SiON layer has a slight excess of nitrogen compared with oxygen, i.e. the relative oxygen ratio R_{Ox}, defined as

$$R_{Ox} = \frac{c_O}{c_O + c_N} \tag{2.1}$$

with c as the atomic concentration of the corresponding element, is about 0.46. On the contrary, the ratio between Si and the elements oxygen and nitrogen is stoichiometric. The SiON compaction is clearly visible due to the reduction of the SiON layer thickness from the original 100 nm down to ~80 nm. Finally, it should be noted that the deposition of Si_3N_4 and SiON on SiO_2 can lead to Si excess at the SiO_2–SiON interface [181].

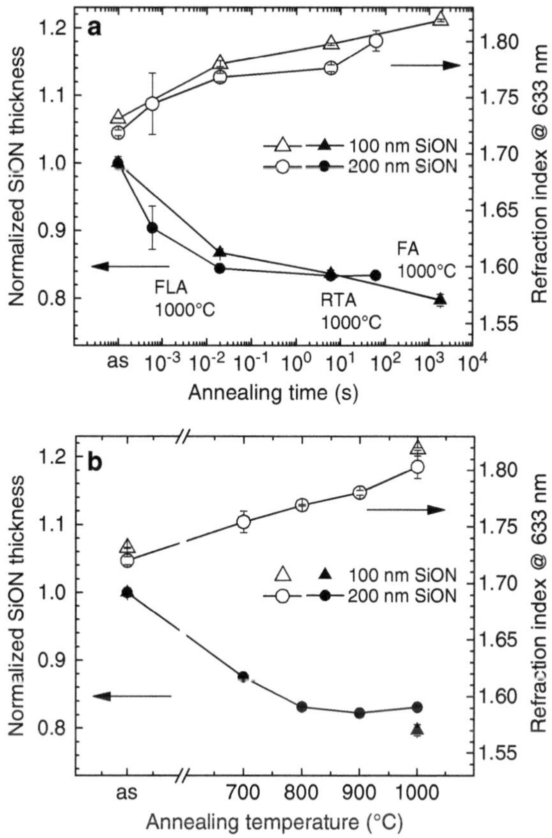

Fig. 2.8 Relative thickness and the refraction index of the SiON layer measured by ellipsometry at 633 nm for different annealing times (**a**) and different annealing temperatures (**b**). The as-deposited SiON is labelled with "as" [180]

2.2.3 Diffusion of Rare Earth Atoms in SiO_2

The redistribution of implanted RE elements under high temperature annealing was investigated by RBS, for the elements Eu, Gd, Tb and Er. These results clearly split the elements into two groups with Eu being in the first and the other elements being in the second group. For the investigated RE elements except Eu, there is little diffusion of RE atoms up to an annealing temperature of 1,000°C. Only for temperatures above 1,000°C, a considerable diffusion of RE atoms towards the interfaces of the SiO_2 layer (Fig. 2.10) can be observed. In addition, the investigation of Gd- and Er-implanted SiO_2 layers reveal only small differences between the RE profiles of the as-implanted state and devices annealed with FA 1,000°C (not shown).

As shown in Fig. 2.11, the picture changes in case of Eu. The implanted Eu profiles after FA 800°C and FLA 1,000°C look very similar to the as-implanted profile, while at FA 900°C or RTA 1,000°C 6 s, first agglomerations of Eu are visible at the SiO_2–SiON interface. Eu segregations at the Si–SiO_2 interface are not yet visible, but with increasing annealing temperature and/or annealing time, the Eu diffusion

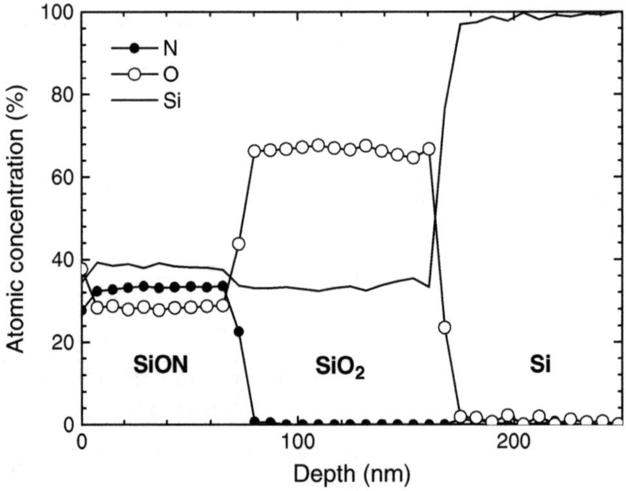

Fig. 2.9 Elemental composition of an unimplanted MOSLED annealed with RTA 1,000°C 6 s, as calculated from depth-dependent AES spectra obtained by sputtering

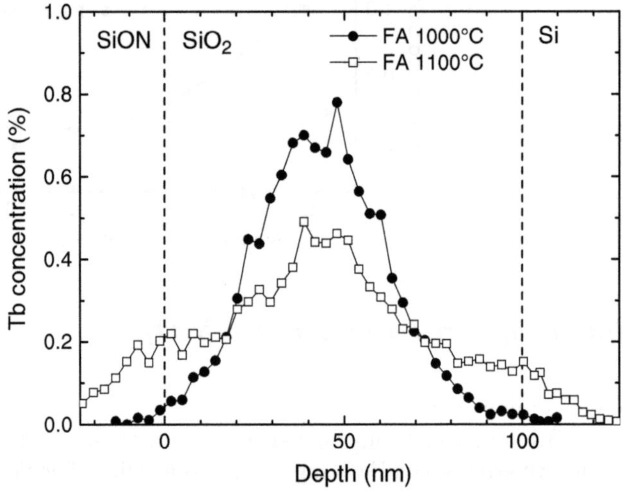

Fig. 2.10 Elemental depth profiles of Tb-implanted MOSLEDs, as calculated from RBS spectra. The implanted Tb dose from RBS was estimated to be about 2.2×10^{15} cm^{-2}

becomes more intense, and at FA 1,000°C, the amount of Eu at both interfaces is higher than in the middle of the oxide layer. If the middle of the oxide layer is defined as the region between 20 and 80 nm of the depth scale, as in Fig. 2.11, only 34% of the implanted Eu can be found within this region. This is even less than in the case of Tb implantation annealed by FA 1,100°C, where the central region still contains about 60% of the implanted Tb dose.

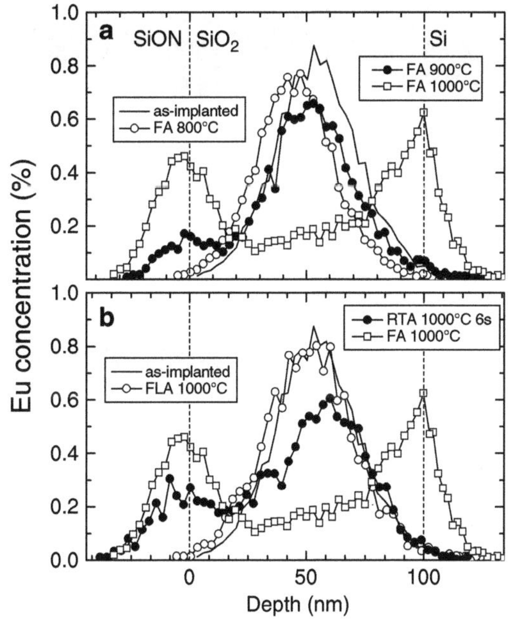

Fig. 2.11 Elemental depth profiles calculated from RBS spectra for Eu in SiO_2 layers implanted with a nominal dose of $3 \times 10^{15}\,cm^{-2}$ and annealed with different annealing temperatures (**a**) and different annealing times (**b**). An atomic density of $6.6 \times 10^{22}\,cm^{-3}$ has been assumed for the SiO_2 layer, to calculate the depth scale (after [132])

The RBS data imply that the diffusion of Eu is strongly enhanced compared with other RE elements. Unfortunately, to the best of our knowledge, no diffusion data of RE atoms in thermally grown SiO_2 can be found in the literature.

2.3 Cluster Formation and Growth

2.3.1 Morphology and Size Distribution

NC formation is a widespread phenomenon for material systems in which impurity atoms were incorporated in concentrations exceeding the solubility limit by far and which were subjected to annealing. Being related to ion-implanted MOSLEDs, the NC formation was investigated by TEM for Si- [182], Ge- [67, 183, 184] and Sn-implanted [185] SiO_2 layers. There are a few reports about clustering in RE-implanted SiO_2, with the majority dealing with Er [186]. According to the present state of knowledge, the cluster formation starts with the formation of ODCs as a result of the nuclear energy deposition during implantation (Sect. 2.2.1) and the deficiency of oxygen introduced by the RE ions. These ODCs serve as seeds for the formation of small clusters by nucleation, which can grow later in size by Ostwald ripening. If the thermal budget applied by annealing is sufficient, one or more cluster bands may arise, depending on the specific implantation and annealing conditions. A single implantation into the middle of the SiO_2 layer usually leads to

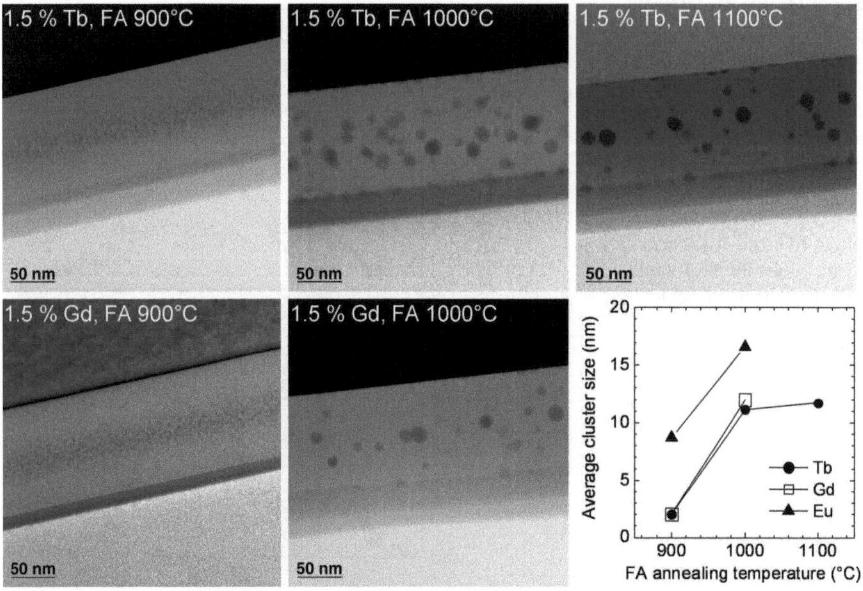

Fig. 2.12 TEM images from a SiO_2 layer single implanted with 1.5% of Tb and Gd furnace annealed between 900°C and 1,100°C (after [187]). The average cluster sizes are given in the *lower right corner*

the formation of a central cluster band with large and small clusters in the centre and the wings of the cluster band, respectively, whereas a double implantation can result in a broad cluster band with a narrow cluster size distribution [67]. Further morphologic features comprise the formation of a small and well-separated cluster band at the Si–SiO_2 interface [184], as well as the formation of cluster bands close to the SiO_2 surface as a consequence of the clash of two diffusion fronts [185].

Figure 2.12 displays a sequence of TEM images from MOSLEDs single implanted with Tb and Gd and annealed by FA between 900°C and 1,100°C. Starting from the top of the image, the bulk Si, a 100 nm thick SiO_2 layer containing the NCs, and a SiON layer with a thickness < 50 nm can be recognized. In case of 800°C (not shown), either no cluster (Tb) or very small clusters of 2 nm or less (Gd) can be observed [135]. At 900°C, both for Tb and Gd, small amorphous clusters with a mean size of about 2 nm are visible. This average size strongly increases to values around 11–12 nm if the annealing temperatures rise to 1,000°C, and first RE decorations at the Si–SiO_2 and the SiO_2–SiON interface can be recognized. A further increase of the annealing temperature to 1,100°C causes only a small increase in cluster size, but significantly more Tb is now found at the interfaces. The average cluster sizes for Tb, Gd and Eu (Fig. 2.13) are given in the graph in the lower right corner of Fig. 2.12.

Similar to the deviant RBS results for Eu (Sect. 2.2.3), the evolution of Eu or Eu oxide clusters differs from that of other RE elements too, as illustrated in Fig. 2.13

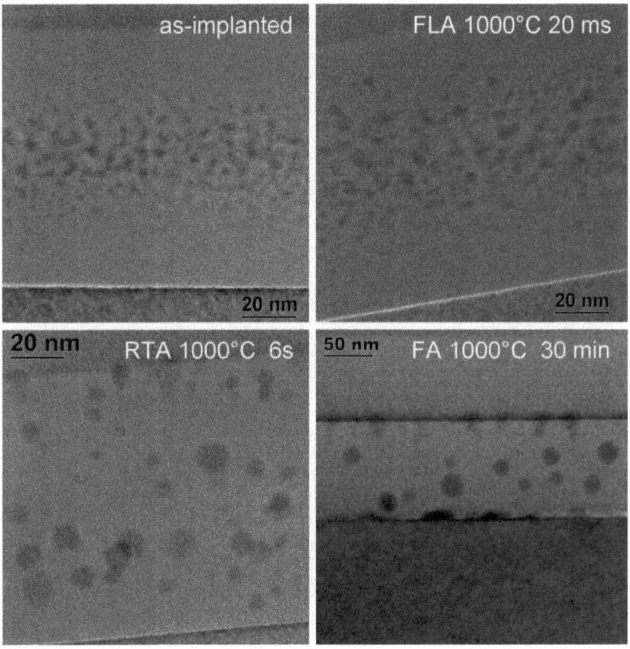

Fig. 2.13 Bright field XTEM images showing the gate oxide layer of Eu-implanted devices, together with its interfaces, to Si (*bottom*) and SiON (*up*). Please note the different magnification for FA 1,000°C (after [132])

showing a sequence of TEM images of Eu-implanted MOSLEDs in the as-implanted state and annealed at 1,000°C with increasing annealing time. The images exhibit the Eu-implanted SiO_2 layer with the adjacent Si substrate and the SiON layer at the bottom and the upper part of the images, respectively. Most noticeably, the cluster formation seems to be greatly accelerated compared with other RE elements. Thus, in the as-implanted state, small amorphous clusters with sizes of around 3–4 nm can already be found in the middle of the SiO_2 layer. As briefly discussed in the preceding chapter, this cluster formation is assumed to be caused by implantation-induced diffusion processes.

Although FLA 1,000°C does not change the picture very much, at RTA 1,000°C 6 s, the formation of larger clusters at the expense of smaller ones caused by Ostwald ripening can be observed. In addition, there is also a significant diffusion of Eu to the interfaces of the SiO_2 layer, which becomes evident by the first Eu segregations at the interfaces and by the broadened cluster band that now extends across the whole oxide layer. In the case of FA 1,000°C, the TEM images show large amorphous clusters up to 20 nm in size and SiO_2 interfaces that are heavily decorated with Eu. At the Si–SiO_2 interface, even the "dissolution" of large clusters can be observed.

A similar development is also found in the case of increasing annealing temperatures. In Fig. 2.14, the development of the cluster size distribution with increasing annealing time at 1,000°C is compared with the cluster development for FA with

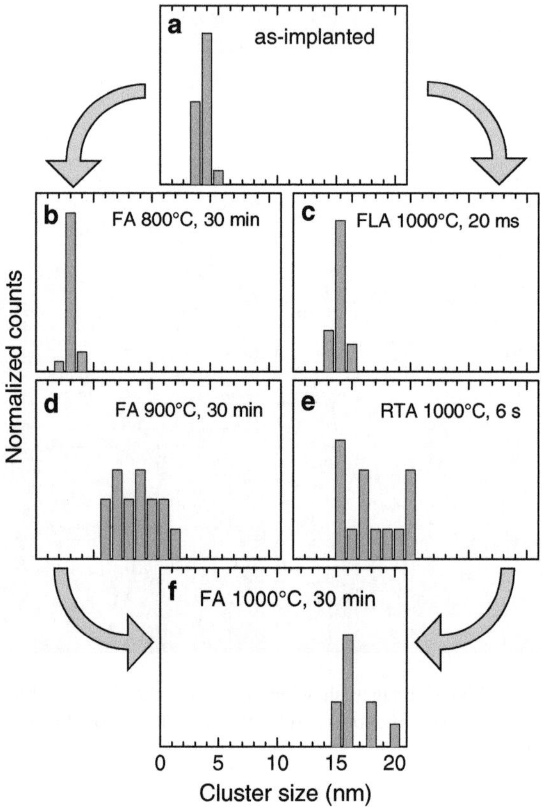

Fig. 2.14 Cluster size distribution of Eu/Eu oxide clusters as a function of the annealing temperature (*left column*) and the annealing time (*right column*). The cluster sizes were determined manually from the XTEM images shown in Fig. 2.13 (after [132])

increasing annealing temperature. Although the cluster sizes were determined manually from the TEM images, resulting in an uncertainty of 1–2 nm, the general tendency is obvious. If the thermal budget exceeds a certain limit either by high temperature or long annealing times, a strong Eu diffusion starts, which finally ends up with the structure known for FA 1,000°C with large Eu clusters and strong Eu segregations at the interfaces. The transition points, namely FA 900°C and RTA 1,000°C 6 s, appear to be equivalent. In the case of FA, the average cluster size is comparable with that of Tb and Gd-implanted devices in Fig. 2.12.

In summary, Eu seems to differ primarily from other RE elements, namely Gd, Tb and Er, because of its accelerated microstructural development, which is possibly caused by a higher diffusion in SiO_2. Consequently, other RE elements pass through the same development stages of morphology as Eu, but at higher thermal budgets. A strong RE diffusion towards the interfaces occurs even at 1,000°C for Eu, whereas at least 1,100°C is needed for the other RE elements (Figs. 2.10 and 2.11). In addition, the cluster sizes for Eu and FA 900°C seem to be similar to those

Table 2.1 Formation enthalpies of different RE oxides [188] and of amorphous SiO_2 at room temperature [189]

	Formation enthalpy (kJ mol^{-1})	Normalized formation enthalpy per reactant (kJ mol^{-1})
Eu_2O_3	−1,651.4	−825.7
Gd_2O_3	−1,819.6	−909.8
Tb_2O_3	−1,865.2	−932.6
Er_2O_3	−1,897.9	−948.95
SiO_2	−910	−910

of Tb and Gd at 1,000°C. However, although the TEM images and the RBS data corroborate the idea of an equal microstructural development with different velocities, it is not yet clear how far this model extends and if it also applies to the structure and composition of different RE clusters.

2.3.2 The Oxidation State of Rare Earth Clusters

RE elements are known to have the strong tendency to achieve a trivalent ionic configuration and to form an oxide, most probably RE_2O_3. This is illustrated in Table 2.1 showing the formation enthalpies of the corresponding RE oxides whose values are in the same order of magnitude than the SiO_2 value, if normalized to the reactant (one RE or one Si atom). As mentioned in Sect. 2.2.1, Si and oxygen are released from the SiO_2 network during the implantation. In the following phase of annealing (and even partly during the implantation), oxygen can be either reintegrated into the SiO_2 network or used to oxidize the RE atoms. There is a competition for oxygen between Si and the RE elements, and as a result the RE atoms will cluster as REO_x, with x depending on the supply and demand of oxygen. As the normalized formation enthalpies of RE_2O_3 except Eu_2O_3 are more negative (more exothermic) than that of SiO_2, and because of the higher demand on oxygen per Si atom, the available oxygen is favourably used to oxidize RE atoms.

At present, there are only a limited number of experimental indications for this scenario, which was described initially for Eu in [190]. In the rare case where crystalline clusters are observed in the TEM images of Eu-implanted MOSLEDs, the inter-planar spacing obtained by Fourier-transformation of the atomic planes of such crystallites confirms the formation of the Eu_2O_3 phase [132]. As the formation enthalpies of Gd_2O_3, Tb_2O_3 and Er_2O_3 are even more negative (more exothermic) than that of Eu_2O_3, these RE elements are also supposed to cluster in an oxidic form. The X-ray absorption fine structure analysis of Tb-implanted SiO_2 layers revealed that Tb is coordinated with either two or six oxygen atoms, and that the latter configuration is favoured by FA 900°C 30 min [191]. However, sixfold oxygen coordination strongly indicates a RE^{3+} configuration, and after FA 900°C 30 min, a major part of the Tb is bound in small clusters. In the case of Ce-rich SiO_2, it was reported that FA 1,200°C 3 h leads to the formation of $Ce_2Si_2O_7$ and related compounds [192]. A similar behaviour was observed for Pr, which formed $Pr_2Si_2O_7$ in Pr-rich SiO_2 after FA 1,100°C 1 h [193].

Chapter 3
Electrical Properties

The MOS system in the form of a Si substrate with a thermally grown SiO_2 layer and a metal gate is one of the best investigated material systems worldwide with early investigations dating back to the 1950s. The knowledge about this material system was extended in the 1960s and 1970s, and later developments perfected the picture with ultrathin SiO_2 layers, MNOS devices and the introduction of high-κ dielectrics. There are a couple of extended and excellent surveys dealing with both the fundamentals of semiconductor physics and the advanced investigation techniques developed and refined over decades, for example, [194–196]. Therefore, it is assumed within the frame of this book that the basic concepts of semiconductor physics like the bandgap model or doping are known, and advanced methods will only be treated in detail as far as they are needed for the understanding of the physical background. Consequently, Sect. 3.1 briefly reviews what is known about the charge injection and transport in MOS devices with respect to the specific structure we have used. Whereas the electrical properties of unimplanted and RE-implanted MOSLEDs in the as-fabricated state are discussed in Sect. 3.2, Sect. 3.3 focuses on these devices under constant current injection. Based on these results, a more general model of the electrical charge trapping behaviour is outlined in Sect. 3.4.

3.1 Charge Injection and Transport in Insulators

3.1.1 The MOS Structure under Fowler–Nordheim Injection

The electrical operation of an MOS device with the intention of light emission requires the flux of charge carriers through the structure. Because of the much higher mobility, electrons are the dominant type of charge carriers, although holes and mobile ions can alter the electric properties considerably. Figure 3.1 shows a basic scheme of an unbiased MOS structure with an additional SiON layer as usually discussed throughout the book. The data of bandgap and band offset energies for Si and SiO_2, SiON and ITO are extracted from [194–200], respectively. For Si, an n-doped wafer with a donator concentration of about $10^{15}\,cm^{-3}$ was assumed leading to a

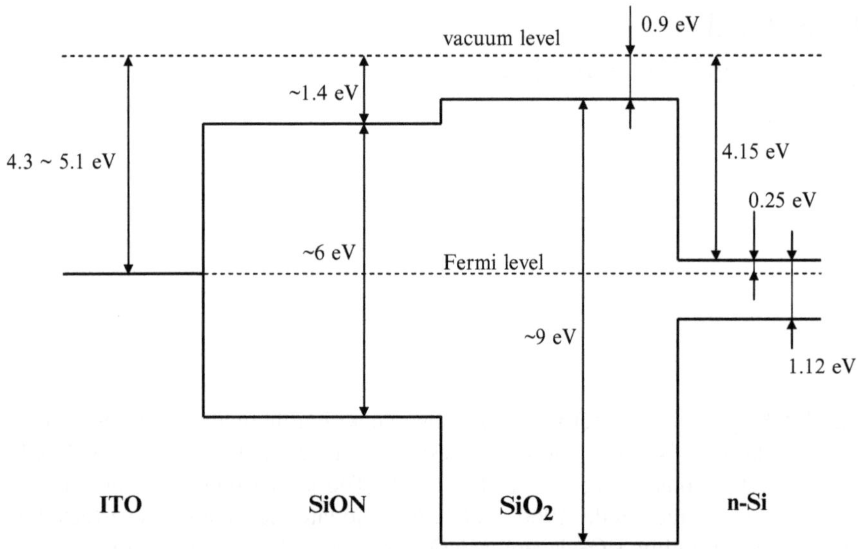

Fig. 3.1 Basic scheme of an unbiased MOS structure with an additional SiON layer

Fermi level position approximately 0.25 eV below the CB edge. The bandgap data for SiON differ slightly in literature as in the region around an O:N ratio of one it sensitively depends on this ratio. Finally, the work function of ITO varies depending on the exact composition [200], the annealing history and even the surface treatment [199]. For the sake of simplification, the band diagram of ITO was plotted as a simple metal although it has a bandgap with a CB partly filled with electrons by tin doping [201].

Generally, the current through an insulator can be either bulk or injection limited. In the present case, the current is clearly injection limited because of the high barrier which electrons have to overcome in order to enter the CB of SiO_2 from the Si substrate. Moreover, at RT and with layer thicknesses in the order of 100 nm, neither thermionic currents (Schottky currents) nor direct tunnelling play any significant role. Charge injection is only possible if a sufficient high voltage is applied which bends down the CB of SiO_2 as illustrated in Fig. 3.2. In such a case, electrons in the CB of Si can tunnel through the approximately triangular barrier into the CB of SiO_2 (process 1). This type of injection is widely known as Fowler–Nordheim (FN) injection [196] and can be assisted by trap-assisted tunnelling (process 2) if there is a sufficient high number of traps close to the Si–SiO_2 interface. In such a case, an electron tunnels first from the CB of Si to a suitable trap site followed by a second tunnelling process from the trap into the CB of SiO_2. This might especially be the case if RE ions diffuse towards the Si–SiO_2 interface under high temperature annealing (see Sect. 2.2.3).

If an electron is once in the CB of SiO_2, there are several possibilities of being transported. If the concentration of defects causing shallow defect levels in

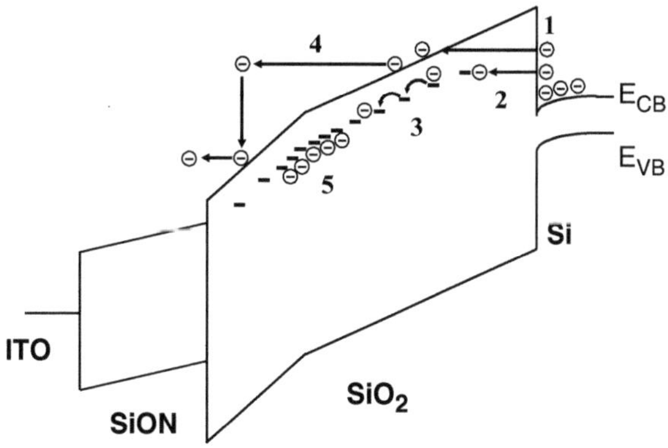

Fig. 3.2 Basic scheme of charge injection and transport phenomena under forward bias. The bandgaps, band offsets and energy distances are not in scale. The individual processes are explained in the text

the bandgap of SiO_2 is high enough, a considerable amount of electrons can be transported via hopping or Poole–Frenkel conduction (process 3). In this case, the electrons move from trap to trap without gaining a noteworthy amount of kinetic energy. This type of conduction is assumed to be dominant in case of SiON or Si_3N_4, but less probably in thermally grown SiO_2 with its low defect density. In this case, electrons can also be transported by the quasi-free movement in the CB of SiO_2 (process 4). The electrons gain kinetic energy when accelerated in the electric field and lose this energy partly and discontinuously by inelastic scattering. These hot electrons are responsible for the excitation of LCs, which is why their properties are discussed in more detail in the following section. In addition, electrons can be trapped at defects in the bulk of the SiO_2 leading to a bending up of the SiO_2 CB, which in turn hinders the further injection of electrons (process 5).

Under the assumption of a triangular barrier, which neglects additional effects like image forces, the injection current density j for FN tunnelling can be written as

$$j = \frac{q^3}{8\pi h}\frac{E^2}{\phi_0}\exp\left(-\frac{2\alpha^*}{3E}\cdot\frac{\phi_0^{3/2}}{q^{3/2}}\right) \quad \text{with } \alpha^* = \alpha\left(\frac{m^*}{m_0}\right)^{1/2} \tag{3.1}$$

and q, E, ϕ_0, m^*, m_0 being the elementary charge, the electric field, the barrier height, the effective and the free mass of the electron, respectively [196]. α is a constant which is given by

$$\alpha = \frac{2\,(2m_0q)^{1/2}}{\hbar} = 1.024\cdot 10^{10}\left[\frac{\sqrt{As\,kg}}{Js}\right]. \tag{3.2}$$

In most cases, $m^* \approx 0.5\,m_0$ is a good approximation; e.g. in [202] a value of 0.42 was used. Figure 3.3a gives an ideal FN characteristic with $\phi_0 = 3.1\,\text{eV}$. To be more realistic, a background current of $10^{-9}\,\text{A cm}^{-2}$ and a virtual electrical BD at $11\,\text{MV cm}^{-1}$ were added. For practical purposes, it is sometimes more comfortable to insert the data in a FN plot where the term $\ln(j/E^2)$ is plotted vs. $1/E$. As seen

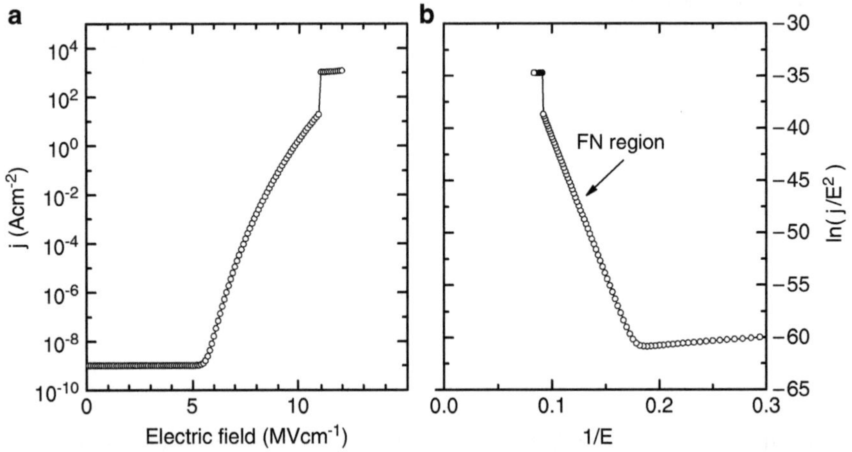

Fig. 3.3 Ideal *IV* characteristic under FN tunnelling with $\phi_0 = 3.1\,\text{eV}$, a virtual background current of about $10^{-9}\,\text{A}\,\text{cm}^{-2}$ and an arbitrary electrical BD at $11\,\text{MV}\,\text{cm}^{-1}$ (**a**). The same data are plotted in a FN plot (**b**)

in Fig. 3.3b, the region of FN tunnelling gives a declining straight line in this plot, for which reason fitting in such a plot is easier.

However, the ideal FN characteristic hardly fits with measured *IV* characteristics. First of all, the existence of a SiON protection layer causes a division of the applied voltage between the SiO_2 layer and the SiON layer. Second, charging is a common phenomenon also leading to a shift of the *IV* characteristic depending on the value and the polarity of the charge. Both effects require corrections which are briefly discussed in Sect. 3.1.3. In addition, FN tunnelling mixes with trap-assisted tunnelling in case of charge traps close to the Si–SiO_2 interface. RE segregations at this interface may have the same effect. As the results of these two types of charge injection are very similar, it is rather difficult to separate their contributions to a real *IV* characteristic. However, in most cases, it is reasonable to fit such an *IV* characteristic with FN and an effective barrier height which implicitly contains the effect of trap-assisted tunnelling.

Finally, a few words about bulk limited currents. This case requires that there is either no injection barrier for charge carriers (known as an ohmic contact) or that the barrier is low enough to allow the injection of a sufficiently high number of charge carriers. Under such conditions, the current is limited by the availability of electronic states in the CB of SiO_2 and/or by space charges which build up during injection. In the present case, the barrier height is usually too high even under high electric fields to shift the limiting current mechanism from injection limited to bulk limited. Nevertheless, it cannot be excluded that space charge limited currents could make a significant contribution in the case of extremely high injection currents just before breakdown; but in practical cases, they will not be considered. Further details can be found in [196, 203].

3.1.2 Hot Electrons in SiO_2

The present picture of the behaviour of electrons in the CB of SiO_2 was developed at the beginning of the eighties [204]. Accordingly, the motion of electrons is mainly controlled by their interaction with the lattice mediated by phonon scattering [205, 206]. This phonon scattering can be divided into a polar mode (LO phonons) and a non-polar mode (TA, LA, TO phonons) whose properties are quite different. In SiO_2, there are two types of LO phonons with energies at 63 and 153 meV which couple efficiently to electrons at low energies with small scattering angles [205, 206]. For electron energies above the phonon energies, the coupling strength between electrons and phonons decreases which leads to a runaway of the kinetic energy without further stabilizing processes. However, the scattering rate of non-polar phonon modes increases strongly with energies above 3 eV, which results in a stable equilibrium distribution of the kinetic energy of hot electrons in SiO_2. The scattering rate of electrons for the different phonon modes as simulated by Fischetti et al. [206] is given in Fig. 3.4.

The consequences to the energy distribution of the hot electrons are the following: The runaway of the kinetic energy occurs already at relatively low electric fields, namely, around 1.5 MV cm^{-1} [205], but is balanced by non-polar phonon scattering when an electric field of 4 MV cm^{-1} is exceeded (Fig. 3.5). Above this field, the average energy slowly increases from 3 to 6 eV for electric fields as large as 16 MV cm^{-1}.

The average energy is dependent not only on the electric field but also on the location in the SiO_2 layer and thus from the SiO_2 layer thickness if it becomes small enough. Clearly, after injection, the electrons need a certain acceleration distance in order to reach their equilibrium stage. In Fig. 3.6, the calculated average energy of injected electrons is displayed as a function of the distance from the injecting interface for an electric field of 4 MV cm^{-1} [206]. When starting with no kinetic energy, the average energy in equilibrium is reached after 10–15 nm, whereby this value slightly increases or decreases with increasing or decreasing electric field,

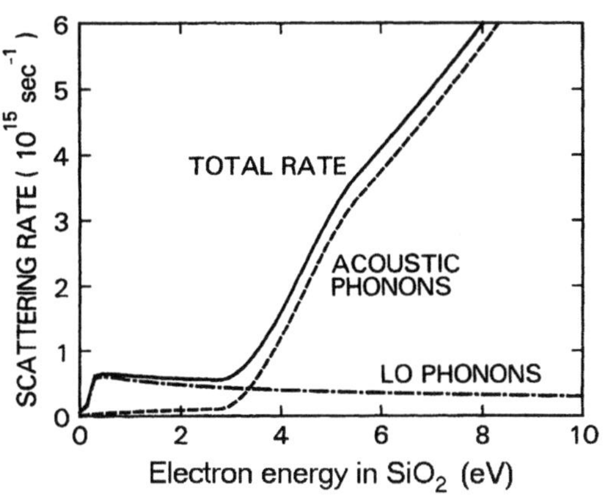

Fig. 3.4 The scattering rate of electrons in SiO_2 for the different phonon modes as simulated by Fischetti et al. [206]

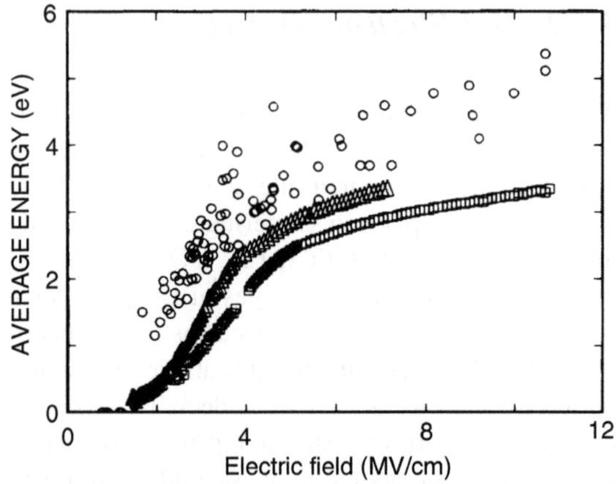

Fig. 3.5 Average energy of hot electrons in SiO_2 as a function of the electric field as determined by vacuum emission (*open circles*) or carrier separation (*all other*) (after [232])

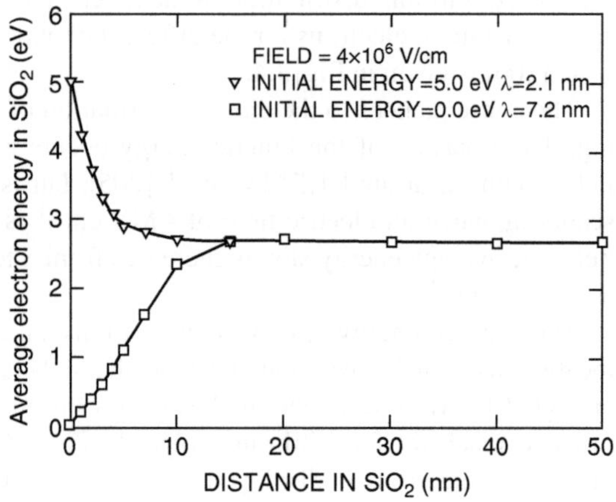

Fig. 3.6 Average energy of electrons injected into SiO_2 as a function of the distance from the injecting interface (after [206]). λ denotes the energy relaxation length in which the difference between the initial kinetic energy and the equilibrium energy has reduced to $1/e(\approx 0.37)$

respectively. Interestingly, deceleration of electrons entering the SiO_2 layer with a larger kinetic energy is even more efficient than acceleration due to the high non-polar phonon scattering rate at high energies. As a consequence, the average electron energy in equilibrium is not reached in SiO_2 layers with a thickness around or below the value of $\sim$15 nm.

As known from vacuum emission measurements [207], the equilibrium energy distribution of the hot electrons is broad with a maximum around the average energy. Whereas there are little changes in the main part of the distribution for a SiO_2 thickness larger than 10 nm, the high energy tail changes more considerably with increasing electric field and increasing SiO_2 thickness. As already mentioned, the average energy increases slightly with increasing electric field, but the high energy tail enhances disproportionately high and extends rapidly to higher energies. Moreover, there is experimental evidence for a small fraction of electrons which face only little phonon scattering. These electrons can gain kinetic energies exceeding by far the bandgap energy of SiO_2 of around 9 eV if the electric field is high and the SiO_2 layer thick enough. In [207], it was found that for 150-nm thick SiO_2

layers and an electric field of 9 MV cm^{-1}, there are electrons whose energy even exceeds a value of 40 eV! Consequently, the damage which hot electrons may produce by impact excitation also depends on the location in the SiO_2 layer. Similar to the discussion about the average energy of the hot electrons, there is a so-called dark space in the order of 30 nm behind the injecting interface where the electrons have no enough energy to create new defects by impact excitation [208]. However, the impact of electrons with such extremely high energies on the stability of the SiO_2 layers seems to be limited as the electric BD field is not drastically decreased at higher oxide thicknesses. In fact, careful investigations revealed that there is a moderate decrease of the QBDvalue with increasing oxide thickness, implying that the contribution of the extremely hot electrons to oxide degradation is significant but not dominating [209, 210].

3.1.3 Charge Trapping and De-trapping in MNOS Systems

Charge trapping and de-trapping are complex and multi-faceted phenomena which depend on plenty of fabrication and operation parameters. This is illustrated by the fact that during the lifetime of a MOSLED, charges in the range of 10^{18}–10^{21} $e^- cm^{-2}$ will cross the oxide layer, but trapped charges with concentrations as low as 10^{12} $e^- cm^{-2}$ already cause considerable shifts of the *IV* and *CV* characteristics and thus of the electrical operation conditions. The main parameters to characterize traps are the trapping cross section, the trap density, the energetic level in the bandgap and the occupancy, that is, which percentage of the traps is already filled by trapped charge carriers. The basic behaviour of traps under charge injection can be described by a first-order kinetics with the assumption that the trapping efficiency – the first derivation of the trapped charge with respect to time – is proportional to the excitation cross section and the number of empty traps. Thus, the dependence of the total trapped charge Q_t can be described as a function of the injected charge Q_{inj} by the equation [211, 212]

$$Q_t = Q_t^{max} \left[1 - \exp\left(-\sigma_i \frac{Q_{inj}}{q} \right) \right], \tag{3.3}$$

with σ_i, Q_t^{max} and q being the trapping cross section, the maximum trapped charge and the elementary charge, respectively. Q_t and Q_t^{max} can be easily replaced by the number of occupied and the total number of traps, respectively. In addition, $Q_{inj} = jt$ holds under a constant injection current j. For the fitting of experimental data, the trapping efficiency

$$P = \frac{dQ_t}{dQ_{inj}} = \frac{\sigma_i}{q} Q_t^{max} \exp\left(-\sigma_i \frac{Q_{inj}}{q} \right) \tag{3.4}$$

is more suitable, as in the plot $\ln(P)$ vs. Q_{inj} the cross sections of different traps can be directly obtained from the slope of the corresponding linear fits.

Depending on whether the trap is negatively charged, neutral, or positively charged, the trapping cross section for electrons can cover the range from 10^{-21} to $10^{-13}\,cm^2$. Large trapping cross sections are difficult to measure (or require very low injection currents) as already a small injected charge is sufficient to fill the traps, which may happen in the ms-range under normal operating conditions. Therefore, the term "unstressed" used to designate freshly prepared MOSLEDs implies that the device has already experienced an injected charge in the order of 10^{14}–$10^{15}\,e^-\,cm^{-2}$. At the other end of the scale, it is also difficult to measure very small capture cross sections as in this case, high injection currents have to be applied for hours to induce a significant trap filling. In addition, the measurement might not be successful because of the unpredictable occurrence of dielectric breakdowns at high injected charges.

The effect of trapped charges can be determined by solving the one-dimensional Maxwell equation

$$\frac{\partial}{\partial x} D(x) = \rho(x), \tag{3.5}$$

with D and ρ, respectively, being the dielectric displacement and the trapped charge distribution in the MOSLED. The approach is straightforward but longsome and can be studied in detail in [196]. The final result is

$$\begin{aligned} - d_{\text{eff}} E_{\text{Ox}}(0) = V_0 &+ \frac{Q_{\text{Ox}}}{\varepsilon_0 \varepsilon_{\text{Ox}}} \left[d_{\text{Ox}} - \overline{x}_{\text{Ox}} \right] + \frac{Q_{\text{N}}}{\varepsilon_0 \varepsilon_{\text{N}}} \left[d_{\text{Ox}} + d_{\text{N}} - \overline{x}_{\text{N}} \right] \\ &+ \frac{d_{\text{N}}}{\varepsilon_0 \varepsilon_{\text{N}}} \left(Q_{\text{ON}} + Q_{\text{Ox}} \right), \end{aligned} \tag{3.6}$$

with

$$d_{\text{eff}} = d_{\text{Ox}} + \frac{d_{\text{N}}}{\varepsilon_{\text{N}}} \varepsilon_{\text{Ox}}. \tag{3.7}$$

Distances are measured with respect to the injecting interface. Whereas the physical quantities Q, ε, d and $\overline{x}$ denote the charge per area, the relative dielectric constant, the layer thickness and the charge centroid, the indices *Ox*, *N* and *ON* mark the assignment of the corresponding quantity to the SiO_2 layer, SiON layer or SiO_2–SiON interface, respectively. $E_{\text{Ox}}(0)$, ε_0 and V_0 are the local electric field at the injecting interface, the dielectric constant and the applied voltage, respectively. Depending on how (3.6) is transformed, either the FN formula (3.1) can be corrected for trapped charges (as e.g. done in [213]) or the change of the applied voltage ΔV_{CC} under constant current injection can be translated in a quantitative value of the trapped charge (assuming that the corresponding charge centroids are known). ΔV_{CC} itself is the difference $V_0(t) - V_0(t_0)$ between the gate voltage V_0 at the time t and an initial value at the time t_0. It should be considered that (3.6) does not cover charges being located at the Si–SiO_2 interface and that $\overline{x}_{\text{Ox}}$ must be larger than the tunnelling distance through the triangular injection barrier.

As the ΔV_{CC} and the *CV* measurement have a different sensitivity to charges at the Si–SiO_2 interface, the independent measurement of ΔV_{CC} and the shift of the flatband voltage ΔV_{FB} allows a separation between the charge Q_{Ox} trapped in the bulk of the SiO_2 layer during constant current injection and a charge Q_{IF} located at the Si–SiO_2 interface. If Q_N and Q_{ON} are negligible, formula [(3.4)–(3.6)] directly leads to

$$\Delta V_{CC} = -\frac{Q_{Ox}}{\varepsilon_{Ox}\varepsilon_0}\left[d_{eff} - \overline{x}_{Ox}\right]. \tag{3.8}$$

The *CV* measurement is sensitive to charges which are located both in the bulk and at the Si–SiO_2 interface and which cause a shift of the *CV* characteristic by a value similar to ΔV_{CC} in formula (3.8) [195]. Considering also Q_{IF}, ΔV_{FB} can be written as [214]

$$\Delta V_{FB} = -\frac{Q_{Ox}}{\varepsilon_{Ox}\varepsilon_0}\left[d_{eff} - \overline{x}_{Ox}\right] - \frac{Q_{IF}}{\varepsilon_{Ox}\varepsilon_0} d_{eff} = \Delta V_{CC} - \frac{Q_{IF}}{\varepsilon_{Ox}\varepsilon_0} d_{eff}. \tag{3.9}$$

Strictly speaking, the voltage shifts still have to be corrected by the work function difference between Si and the gate material, which is usually in the order of 1 V and below [195]. However, the measured voltage shifts are much higher than that for which reason this contribution can be easily neglected. As in most cases (for low interface trap densities), the shift of the *CV* characteristic is approximately a parallel shift also the shift of the midgap voltage ΔV_{MG} can be used instead of ΔV_{FB}. In addition, an *IV* characteristic also shifts by the value ΔV_{IV} equal to ΔV_{CC} from (3.8) assuming that there is no significant charge trapping *during* the *IV* measurement. Finally, it should be mentioned that it is possible to separate the trapped charge Q from the charge centroid $\overline{x}_{Ox}$ by recording the *IV* characteristics under both polarities and by carefully deducting other effects like the work function difference [196]. A comprehensive introduction into the techniques of charge injection can be found in [211].

3.2 Rare Earth Implanted Unstressed Light Emitters

3.2.1 The Unimplanted State

The discussion about the electrical properties of unstressed RE-implanted MOSLEDs starts with unimplanted structures as already the deposition of SiON followed by annealing gives very different results and leads to a pronounced dependency on the annealing procedure. As already mentioned, the terminus "unstressed" is an idealization because the performance of an electrical measurement requires a minimum current load, and thus a certain number of electrons, which have to be injected into the device. Whereas this current load is minimal in case of *CV*, it is substantial for *IV* and ΔV_{CC} measurements. In case of unstressed devices, charges in the

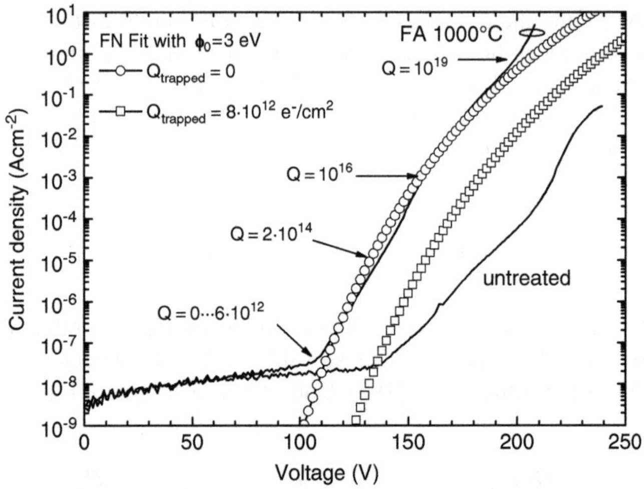

Fig. 3.7 *IV* characteristics of unimplanted devices either untreated (not annealed) or treated with FA 1,000°C. Two FN fits with $\phi_0 = 3\,\text{eV}$ without (*open circles*) and with a trapped negative charge of $Q_{\text{inj}} = 8 \times 10^{12}\,\text{e}^-\,\text{cm}^{-2}$ (*open squares*) are also plotted. Further details are given in the text

bulk of the dielectrics can be deduced from the comparison of experimentally measured *IV* data with ideal FN characteristics. Figure 3.7 shows the *IV* characteristics of unimplanted devices with 100 nm SiO_2 and 100 nm SiON either untreated (not annealed) or treated with FA 1,000°C together with the charge which was already injected into the device when a certain point of the *IV* characteristics is reached.

Starting the discussion with FA 1,000°C, an FN fit (open circles) according to (3.1) with $m^* \approx 0.5m_0$ and $\phi_0 = 3\,\text{eV}$ and an effective layer thickness according to (3.7) was used to model the measured *IV* characteristics. The fit considers no additional charges but a reduced SiON layer thickness of 82 nm according to the compaction of the SiON layer under high temperature annealing (Figs. 3.2–3.8). As seen, the fit reproduces the *IV* characteristic for FA 1,000°C quite reasonable. In contrast to FA 1,000°C, the *IV* characteristic of the untreated device is shifted to higher voltages and runs shallower. Assuming that the SiON deposition itself has little influence on the quality of the Si–SiO_2 interface and thus on the injection barrier height, this shift can be explained by the appearance of negative charges in one of the two dielectric layers. A second fit (open squares) now considers a 100 nm thick SiON layer and the effect of a trapped negative charge of $Q_{\text{inj}} = 8 \times 10^{12}\,\text{e}^-\,\text{cm}^{-2}$ at the SiO_2–SiON interface region according to (3.6). Changes of the trapped charge or the charge centroid shift the FN fit to lower or higher voltages, but further on it fails to reproduce the measured *IV* characteristic and reveals the phenomenon of charge trapping *during* the *IV* measurements. According to this, the difference between the FN fit and the measured *IV* characteristics of the untreated device is caused by a negative charge which is trapped during the measurement and in excess to the negative charge already considered in the FN Fit. Fortunately, this behaviour is restricted to the untreated case; other *IV* characteristics normally lie

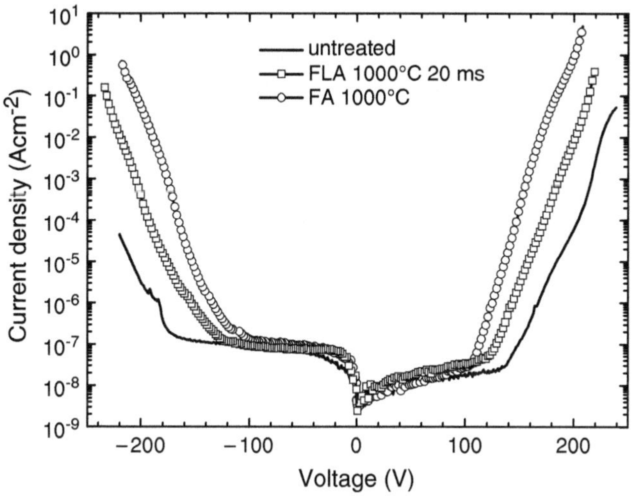

Fig. 3.8 *IV* characteristics of unimplanted devices which are untreated, treated with FLA 1,000°C 20 ms or treated with FA 1,000°C

between FA 1,000°C and the untreated characteristic and seem to be parallel shifted to the first one (Fig. 3.8). Nevertheless, this introduces an additional uncertainty when comparing the shift of *IV* characteristics of different devices, but does not affect the validity of the tendencies deduced from such comparisons. Generally, the shifts of *IV* characteristics were determined at a current density of 4×10^{-5} A cm^{-2}.

Figure 3.8 displays again the *IV* characteristics of unimplanted devices either untreated or treated with FA 1,000°C, but this time in forward and backward direction and together with an unimplanted device treated with FLA 1,000°C 20 ms. The *IV* characteristic for FLA 1,000°C 20 ms lies between that of FA 1,000°C and the untreated one indicating that a smaller negative charge is trapped compared to the untreated case. In Sect. 2.2.2, it was outlined that the hydrogen content strongly decreases in the order of untreated device – FLA 1,000°C 20 ms – FA 1,000°C, which is the same order of the *IV* characteristics shift to lower absolute values of the voltage in Fig. 3.8. Therefore, it seems that hydrogen introduces additional *negative* charges into the SiON layer and/or the SiO_2–SiON interface region leading to the observed shift of the *IV* characteristics.

In literature, however, the incorporation of hydrogen in SiON or Si_3N_4 is usually associated with the occurrence of positive charges [171, 215], for which reason this point requires a more detailed discussion. The topic itself is complex as the magnitude of the positive charge varies with the deposition conditions [216]. In addition, annealing in a hydrogen-rich ambient increases the hydrogen content in the corresponding layers, but decreases the amount of available charge traps by passivation [217]. The hydrogen atom bound to Si, oxygen or nitrogen is supposed to be stable in the neutral and positively charged state, and a couple of reactions for the generation of positively charged bond defects involving hydrogen can be found in [218].

The situation changes if not only the initial charge but also the trapping under charge injection is considered. In their neutral states, O- and N-atoms can trap free carriers of either sign, but the participation of hydrogen is required in order to change the bonding coordination [218]. The K^0 centre $N_3{\equiv}Si\bullet$ can also trap a positive or negative charge resulting in the formation of a K^+ or K^- centre, respectively, and there is evidence for the fact that the formation of a K^+–K^- pair is energetically more favourable than the formation of two neutral K^0 centres [219]. In addition, it was found that electron injection into thin SiON layers causes an electron trapping proportional to the amount of hydrogen in this layer [220].

In order to bring these aspects in agreement with the present experimental results, it can be stated that hydrogen, although it may introduce a positive charge in the very beginning, strongly supports electron trapping if electrons are injected. Such a behaviour was observed in [221] where the implantation of hydrogen into Si_3N_4 caused a negative ΔV_{FB} shift (indicating a positive charge), but a positive shift of the *IV* curve similar to that in Figs. 3.7 and 3.8. Obviously, the small current load at the beginning of an *IV* characteristic is enough to induce such a strong electron trapping in hydrogen-rich layers that they appear to be negatively charged from the beginning.

The *IV* characteristics in backward direction (Fig. 3.8) mirror the behaviour of those in forward direction except that they are additionally shifted to higher absolute values of the voltage (neglecting the polarity). This is most probably due to a higher injection barrier for electrons from ITO into the SiON layer. This behaviour is observed for most of the investigated devices but not for all as the work function of ITO is quite sensitive to the fabrication conditions (Sect. 3.1.1).

In contrast to *IV* measurements, the corresponding high-frequency *CV* measurements (100 kHz) show only small deviations from the ideal *CV* characteristic. The sequence of *CV* characteristics of unimplanted devices with a 100 nm SiO_2 and a 200 nm SiON layer given in Fig. 3.9 exhibits that there is a weak shift to negative voltages for low FA temperatures, implying a weak positive charge which, however, is on the edge of significance as voltage shifts up to 2 V are in the range of statistical deviations between different wafers. In addition, the characteristics are shallower than the ideal *CV* characteristic, indicating that there is a significant concentration of traps at the Si–SiO_2 interface [195].

The annealing dependence of unimplanted devices is summarized in Fig. 3.10 showing both the shift of the *IV* characteristic ΔV_{IV} and that of the flatband voltage ΔV_{FB} relative to the corresponding ideal characteristic (dashed line) as a function of the annealing temperature (a) and the annealing time (b). The SiO_2 layer thickness is always 100 nm, and the untreated device is labelled with "as" on the annealing temperature or time axis. Clearly, the *IV* characteristics strongly shift to lower voltages (and thus lower electric fields) with increasing thermal budgets and reach a saturation value for either FA 800°C or RTA 1,000°C 6 s. It has to be noted that the compaction of the SiON layer due to annealing accounts for approximately 20% of the large *IV* voltage shift of the untreated device. For the comparison of annealed devices among each other, the influence of compaction is negligible. In contrast

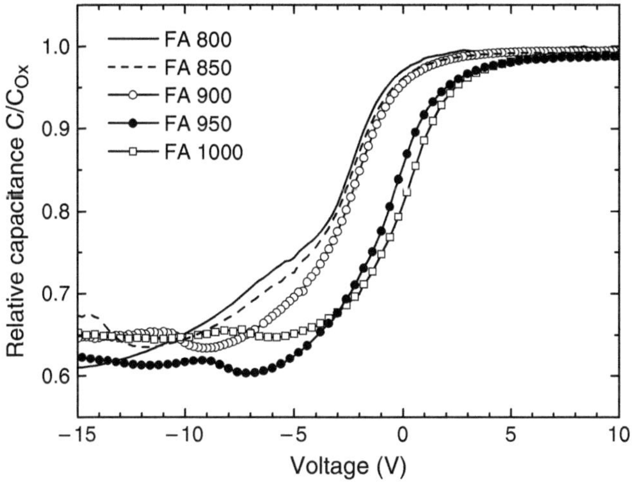

Fig. 3.9 *CV* characteristics of unimplanted devices with a 100 nm SiO_2 and a 200 nm SiON layer for different FA temperatures

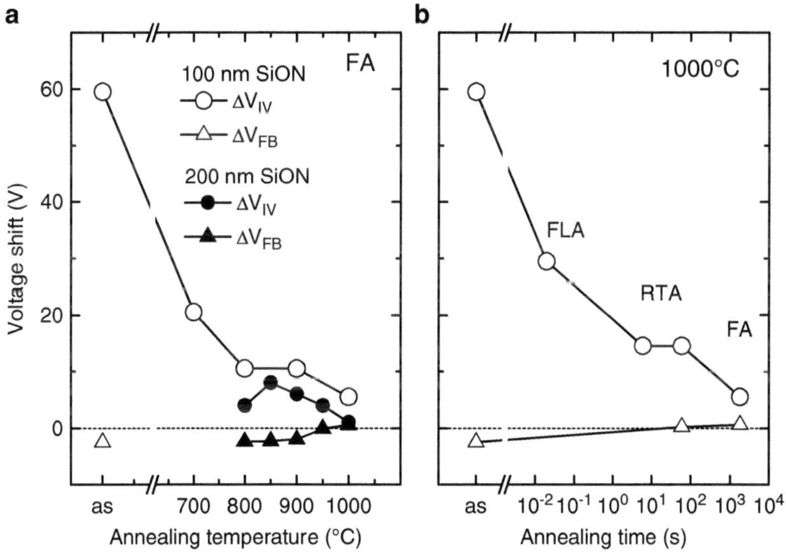

Fig. 3.10 Shift of the *IV* characteristic ΔV_{IV} and the flatband voltage ΔV_{FB} of unimplanted devices in comparison with the corresponding ideal characteristic (*dashed line*) as a function of the annealing temperature (**a**) and the annealing time (**b**)

to *IV*, there is either no shift of the flatband voltage for high thermal budgets or only very small negative values for ΔV_{FB} for low thermal budgets at the edge of significance. This leads to the assumption that the negative charge causing the shift of the *IV* characteristic is outbalanced by positive charges at the Si–SiO_2 interface which is probably done by charging the aforementioned interface traps.

3.2.2 *Annealing Dependence*

As shown, the unimplanted MOSLED exhibits already a strong annealing dependence due to the out-diffusion of hydrogen. RE-implanted devices add now another dependency which is mainly related to the behaviour of implantation-induced defects and the structural changes like diffusion and cluster formation the implanted RE ions will undergo. As this RE-related contribution is in the same order or even smaller than the hydrogen-related part of the annealing dependence, it is not easy to separate these effects from each other. Nevertheless, the main tendencies and the qualitative behaviour can be extracted.

Figure 3.11 displays the difference $\Delta V_{\mathrm{IV}}^{\mathrm{RE}} - \Delta V_{\mathrm{IV}}^{0}$ between the voltage shift $\Delta V_{\mathrm{IV}}^{\mathrm{RE}}$ of the *IV* characteristics of MOSLEDs implanted with 1.5% of a RE element and the corresponding voltage shift $\Delta V_{\mathrm{IV}}^{0}$ of the unimplanted device for various annealing conditions. Starting with the as-implanted state, the implantation of REs shifts the *IV* characteristic to higher voltages, which is interpreted as the creation of a negative charge in addition to the negative charge already induced by the hydrogen load (see previous chapter). The centroid of this additional charge is not exactly known but assumed to be located around R_P. As different RE elements cause a similar voltage shift, the additional negative charge seems to be caused rather by implantation-induced defects than by the specific RE element itself.

In case of Eu, the effect of an additional negative charge even increases for low temperature annealing (FA 700°C), but rapidly decreases for increasing annealing

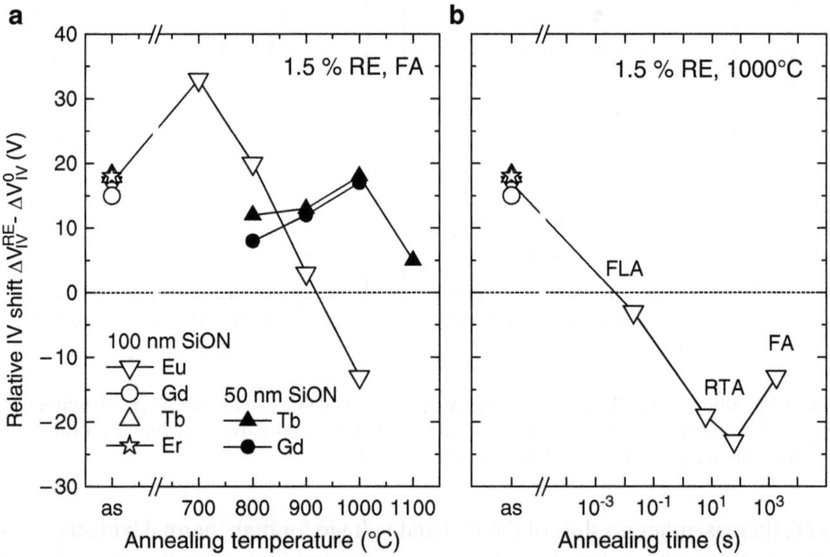

Fig. 3.11 Difference between the voltage shift of the *IV* characteristics of MOSLEDs implanted with 1.5% of a RE element and the corresponding voltage shift of the unimplanted device. Whereas the SiO_2 thickness amounts to 100 nm, the SiON thickness is either 50 or 100 nm

temperatures (Fig. 3.11a). For FA 900°C, the curve crosses the zero line and the effect is turned into the opposite for FA 1,000°C, implying that now positive charges dominate in the dielectric layer. Looking at the annealing time dependence (Fig. 3.11b), the intersection with the zero line is already reached for FLA 1,000°C, and RTA 1,000°C gives results similar to FA 1,000°C. This behaviour can be interpreted as follows: The annealing of implantation-induced defects, especially those induced by the electronic energy deposition during implantation, take place on a relative short time scale, but need a sufficiently high temperature. Therefore, FLA 1,000°C is already able to eliminate an essential part of the defects (not all, as the incorporation of RE ions into the SiO_2 matrix always leaves some structural defects behind), whereas FA 800°C is, regardless of the long anneal time, not enough to do so. The appearance of a positive charge for high thermal budgets can be correlated with the formation of larger REO_x clusters and the onset of a strong RE diffusion towards the interfaces of the SiO_2 layer.

There are less experimental data for other RE elements; most of them originate from devices annealed with FA 900°C which is discussed in the subsequent chapter. Qualitatively, it can be noticed that other RE elements also shift the *IV* characteristic to higher voltages, but there is not such a pronounced annealing dependence as in the case of Eu. A shift to lower voltages even for high temperature annealing was not observed for other RE elements which corresponds well to the fact that certain microstructural features like large REO_x clusters and massive RE agglomerations at the interfaces appear, if at all, only at temperatures as high as 1,100°C. The lower shift for Tb and FA 1,100°C (Fig. 3.11a) might be the onset of such a development.

Figure 3.12 compares the *CV* characteristics of Eu- and Tb-implanted MOSLEDs for different annealing temperatures. In case of Eu, the consequent formation of a positive charge with increasing FA temperature is observed (Fig. 3.12a). However, for annealing temperatures above 900°C, the *CV* characteristics degrade due to Fermi level pinning which results in a decreasing ratio between the maximum and minimum values of the measured capacitances. In contrast to this, the Tb-implanted, unstressed structures show a slight shift of the *CV* characteristics towards positive values of the applied voltage (Fig. 3.12b) which corresponds to a decrease of the positive charge and/or to an increase of the negative charge in the SiO_2. However, an increase of the FA temperature up to 1,100°C results in a strong increase of the positive charge in the dielectric. The comparison of Figs. 3.11 and 3.12 reveals that the *CV* characteristics mirror partly the behaviour of the *IV* characteristics: In case of Eu, an increase of the positive charge (or decrease of a negative one) is visible for the transition FA 800 to FA 900°C, and the same can be recognized for FA 1,000°C to FA 1,100°C in case of Tb. However, there is also a significant difference: Whereas the *IV* characteristics of Eu 900°C and Tb 1,100°C still indicate a negative, albeit small charge, the *CV* characteristics display undoubtedly a larger positive charge which, as not seen by *IV*, is most probably located at the Si–SiO_2 interface.

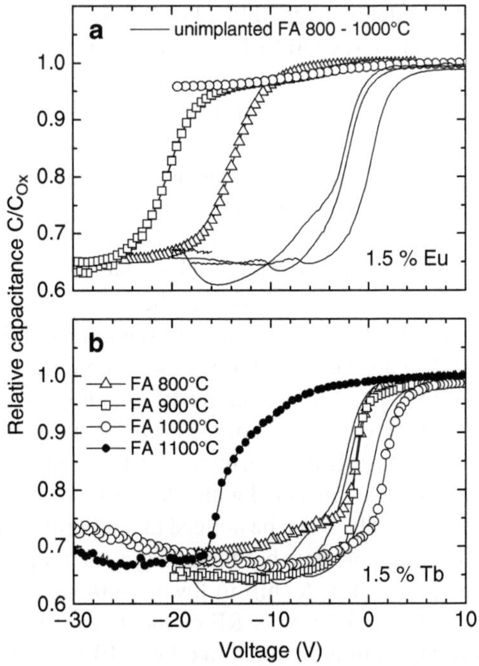

Fig. 3.12 *CV* characteristics of Eu- (**a**) and Tb-implanted MOSLEDs (**b**) for different FA temperatures (after [214])

3.2.3 *Concentration Dependence*

The concentration dependence is as complex as the annealing dependence and closely linked with the latter one. Figure 3.13a shows the difference $\Delta V_{\mathrm{IV}}^{\mathrm{RE}} - \Delta V_{\mathrm{IV}}^{0}$ as a function of RE peak concentration for various RE elements after FA 900°C. In most cases, the shift to higher voltages is the larger the higher the concentration is; most pronounced for Ce and Tb. For high RE concentrations (>3%), the *IV* characteristics start to shift back to smaller voltages which is probably, similar to high temperature annealing, related to the appearance of larger REO_x clusters and RE interface decorations assuming that such features are favoured by high RE concentrations. However, as illustrated in Fig. 3.13b, the situation can be different for other annealing conditions. For RTA 1,000°C 6 s, 0.1 and 0.5% Eu do not differ very much from the unimplanted devices, but for Eu concentrations >1% there is a strong shift to lower voltages with increasing Eu concentration. In contrast to this, for FLA 1,000°C 20 ms, there is a pronounced shift with Eu concentration, whereas for FA 900°C (Fig. 3.13a) and 1,000°C (Fig. 3.13b), all Eu concentrations feature nearly the same voltage shift. In addition, Eu seems to be more sensitive to the annealing conditions than other RE elements.

Figure 3.14 displays the *CV* characteristics of RE-implanted MOSLEDs with a RE peak concentration of 1.5% and annealed at FA 900°C. Whereas Eu shows a large shift to negative voltages implying a positive charge at the Si–SiO_2 interface, the other investigated RE elements deviate only slightly from the *CV* characteristic

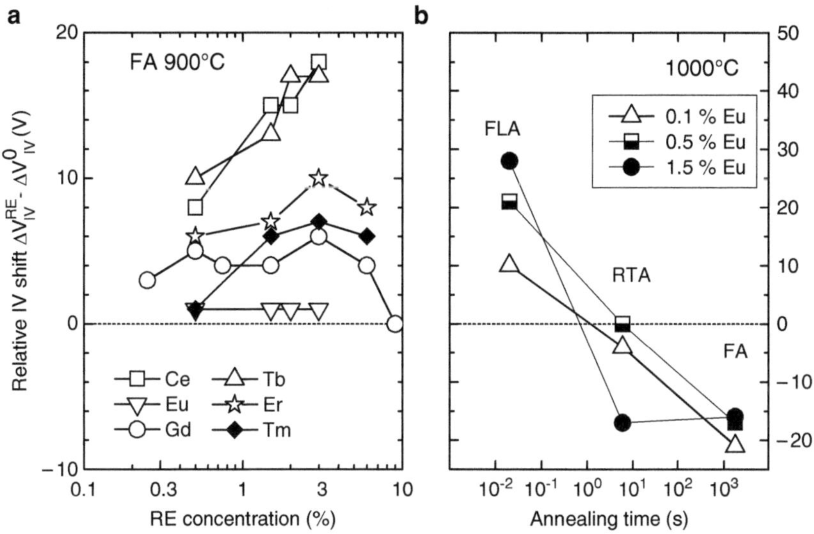

Fig. 3.13 Difference between the *IV* voltage shifts of RE-implanted and unimplanted MOSLEDs as a function of RE concentration for different RE elements (**a**) and different annealing times (**b**). Please note the different scales in (**a**) and (**b**)

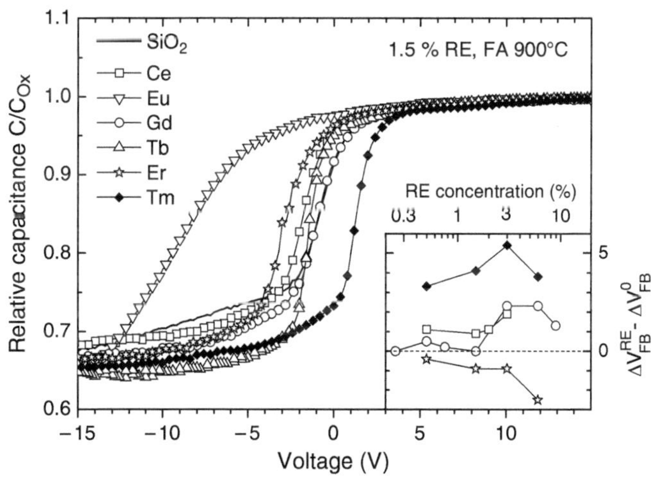

Fig. 3.14 *CV* characteristics of RE-implanted MOSLEDs with a RE peak concentration of 1.5%. The *inset* shows the *CV* voltage shifts of RE-implanted MOSLEDs compared to the unimplanted device as a function of RE concentration. All structures were annealed at FA 900°C

of the corresponding unimplanted device. The concentration dependence of different RE elements, displayed in the inset of Fig. 3.14 as difference between the shift of the flatband voltage of the RE-implanted device ΔV_{FB}^{RE} and that of the unimplanted device ΔV_{FB}^{0}, also does not show a pronounced dependency.

3.3 Rare Earth Implanted Light Emitters Under Constant Current Injection

3.3.1 Low and High Injection Currents

For ΔV_{CC} measurements, a constant current is injected into the devices while the change of the applied voltage is monitored with time and thus with the injected charge. Similar to the *IV* measurements, the change of the applied voltage can give certain information on the trapped charge and how it changes with increasing charge injection. The observed voltage shift also obeys (3.6). However, the value of ΔV_{CC} itself moderately depends on the injection current. Figure 3.15 shows a sequence of

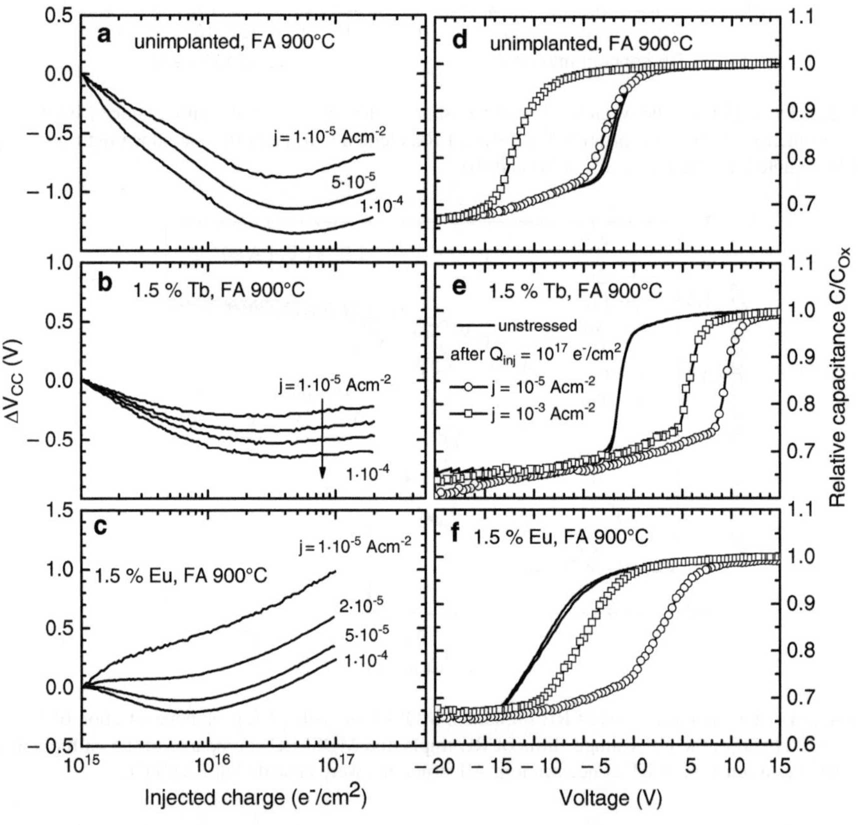

Fig. 3.15 ΔV_{CC} characteristics recorded at different injection current densities for unimplanted (**a**), Tb-implanted (**b**) and Eu-implanted devices (**c**) (after [214]). The *CV* characteristics of the corresponding devices before and after the injection of a charge of about $10^{17}\,e^{-}\,cm^{-2}$, using an injection current density of 10^{-5} or $10^{-3}\,A\,cm^{-2}$, are given to the *right* (**d–f**). All devices were annealed with FA 900°C

ΔV_{CC} characteristics recorded at different levels of the injection current density for an unimplanted (a), a Tb-implanted (b) and an Eu-implanted device (c). Although the characteristics are very similar in shape, they exhibit a systematic shift: with increasing injection current density comparable ΔV_{CC} values become more negative. This can be interpreted either as the reduction of a trapped negative charge or as the buildup of a larger positive charge in the bulk of the dielectric layer system. The same tendency can be observed in *CV* measurements. Figure 3.15d–f compares the *CV* characteristics of unimplanted, Tb- and Eu-implanted devices before and after the injection of a charge of about $10^{17}\,e^{-}\,cm^{-2}$ using an injection current density of 10^{-5} or $10^{-3}\,A\,cm^{-2}$. In all cases, ΔV_{FB} is more negative for the high injection current density of $10^{-3}\,A\,cm^{-2}$. Thus, if the *CV* characteristics of the stressed devices are on the *left* side of the initial, unstressed characteristic, higher injection current densities cause the buildup of a *larger positive* charge (Fig. 3.15d). Correspondingly, higher injection current densities cause the buildup of a *smaller negative* charge (Fig. 3.15e, f) in case the *CV* characteristics of the stressed devices are on the *right* side of the initial, unstressed ones.

The electric field which has to be applied in order to induce a certain injection current increases with increasing injection current densities. There are two processes whose field dependencies have to be considered. At first, the average energy of hot electrons increases with increasing electric fields, and this in turn increases the trap and defect generation rate. The creation of new defects and traps is usually associated with the creation of positive charges by a couple of processes including anode hole injection, trap-to-band and band-to-band impact ionization [222]. Thus, an increasing electric field enhances such processes which are discussed in more detail in Sect. 6.1.1. Second, with increasing electric fields, the balance between trapping and de-trapping shifts to the favour of the latter one resulting in a lowering of the occupancy of electronic traps [222] and thus in a smaller negative charge for higher injection currents.

3.3.2 Charge Trapping in Unimplanted and RE-Implanted Structures

In order to monitor the charge trapping over a wide range of injected charges, ΔV_{CC} and ΔV_{MG} characteristics[1] were recorded using an injection current density of 10^{-5} or $10^{-3}\,A\,cm^{-2}$, defined as low and high current in the following, respectively. Thereby *CV* characteristics were measured after defined periods of charge injection to obtain the individual points of the ΔV_{MG} characteristic. The characteristics are combined in Fig. 3.16 for an unimplanted and a Tb-implanted MOSLED which were both annealed by FA 900°C. The shift of the mid-gap voltage ΔV_{MG} is

[1] The shift of the mid-gap voltage V_{MG} gives more or less the same information as the shift of the flatband voltage V_{FB}. For the exact definition see [250].

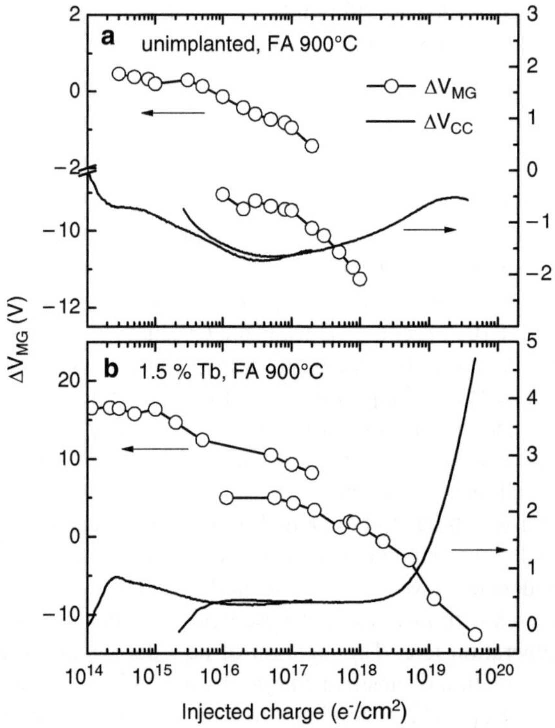

Fig. 3.16 Shift of the constant current voltage ΔV_{CC} (*open circles*, *left scale*) and the shift of the mid-gap voltage ΔV_{MG} of *CV* characteristics (*solid lines*, *right scale*) as a function of the injected charge in unimplanted (**a**) and Tb-implanted MOSLEDs (**b**). Both devices are annealed by FA 900°C. The low and high current characteristics were recorded using an injection current density of 10^{-5} and $10^{-3}\,\mathrm{A\,cm^{-2}}$, respectively. Please note the different scales (after [214])

measured relative to the unstressed device and exhibits, as discussed in the previous chapter, the typical shift of the high current characteristic ($10^{-3}\,\mathrm{A\,cm^{-2}}$) to more negative values (more positive or less negative charges). Whereas the low current ΔV_{CC} characteristic is measured relative to the initial value at about $10^{14}\,\mathrm{e^-\,cm^{-2}}$, the high current ΔV_{CC} characteristic is normalized, that is, parallel shifted on the voltage scale in such a way that the ΔV_{CC} values of both characteristics are equal at $10^{17}\,\mathrm{e^-\,cm^{-2}}$.

For further discussion, the wide range of injected charges can be roughly divided in the ranges of low charge injection (10^{14}–$10^{16}\,\mathrm{e^-\,cm^{-2}}$), medium charge injection (10^{16}–$10^{18}\,\mathrm{e^-\,cm^{-2}}$) and high charge injection ($>10^{18}\,\mathrm{e^-\,cm^{-2}}$). Naturally, traps with a large or small trapping cross section dominate the charge trapping in the range of low and high charge injection, respectively. In the following, the qualitative behaviour of charge trapping in unimplanted and RE-implanted MOSLEDs is presented; quantitative trap parameters were determined and tabulated in [214].

For the unimplanted device (Fig. 3.16a) and low-level charge injection, the buildup of a net positive charge can be traced both in the ΔV_{CC} and in the ΔV_{MG} characteristic. For medium-level charge injection, the ΔV_{CC} characteristic reaches a minimum and starts to rise, whereas the ΔV_{MG} characteristic continues to decrease. For high-level charge injection, a moderate trapping of negative charges or an equivalent decrease of a trapped positive charge can be observed. The parallel or different run of the ΔV_{CC} and ΔV_{MG} characteristics implies that the positive charge is mainly trapped in the bulk of the dielectric layers for low-level charge injection and in the Si–SiO_2 interface region for larger injected charges. A priori the accumulated positive bulk charge is not necessarily located only in the SiO_2 layer. However, the strong similarities between the trapping parameters of the present system with a SiON and an ITO layer and those of the system Al–SiO_2–Si [212] indicate that the charge centroid is most probably located in the SiO_2 layer.

RE implantation alters the picture in some aspects. As shown in Fig. 3.16b at the example of Tb, the shape of the ΔV_{CC} and ΔV_{MG} characteristics is similar for low-level charge injection, but the ΔV_{MG} characteristic is strongly shifted to positive values. Combining this fact with the previous findings, namely, that the *CV* characteristic of an unstressed Tb device is only slightly shifted relative to that of an unimplanted device (Figs. 3.14 and 3.15d, e), the following scenario is probable: The Tb implantation creates defects with such a large trapping cross section for electrons that they are already filled after a charge injection of $10^{14}\,e^-\,cm^{-2}$ or even less. The very first slope in the ΔV_{CC} characteristic, which is characterized by an electron trapping cross section in the order of $10^{-14}\,cm^2$ [214], is an indication of this. For medium and high-level charge injection, the ΔV_{MG} characteristic continuously decreases proving the continuous accumulation of positive charges at the Si–SiO_2 interface, whereas the bulk of the SiO_2 layer seems to be free of charge trapping for more than two orders of magnitude before it faces a strong electron trapping for high-level charge injection.

The situation changes again if the case of Tb, FA 1,100°C and Eu, FA 900°C is considered. Regarding their microstructure, both are characterized by RE agglomerations at the SiO_2 boundaries (Sect. 2.2.3) and REO_x clusters of comparable size (Sect. 2.3.1). Indeed, the charge trapping of Tb, FA 1,100°C has more in common with that of Eu, FA 900°C than that of Tb, FA 900°C. Starting with low-level charge injection, the behaviour of the Eu-implanted device (Fig. 3.17b) is completely different from that of Tb, FA 900°C (Fig. 3.16b): In contrast to Tb, the ΔV_{CC} and ΔV_{MG} characteristics increase with increasing injected charges, indicating the buildup of a negative charge in the bulk of the SiO_2 layer. The behaviour of Tb 1,100°C with a first trapping of negative charges ($<3 \times 10^{14}\,e^-\,cm^{-2}$) followed by positive charge trapping (3×10^{14}–$5 \times 10^{15}\,e^-\,cm^{-2}$) is somewhere between that of Eu, FA 900°C and Tb, FA 900°C.

Already for medium-level charge injection, there is a strong electron trapping both visible in the ΔV_{CC} and ΔV_{MG} characteristics, and this is the most obvious difference to Fig. 3.16. The accumulation of positive charges at the Si–SiO_2 interface is either not present or too weak to balance the influence of the bulk electron trapping on the *CV* characteristic. Only for high-level charge injection, the interface

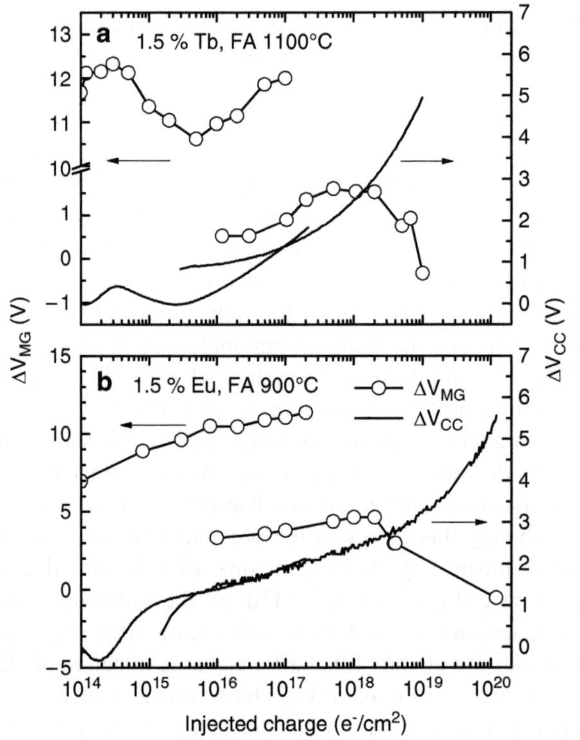

Fig. 3.17 Shift of the constant current voltage ΔV_{CC} (*open circles, left scale*) and the shift of the mid-gap voltage ΔV_{MG} of *CV* characteristics (*solid lines, right scale*) as a function of the injected charge in a Tb-implanted MOSLED annealed by FA 1,100°C (**a**) and an Eu-implanted MOSLED annealed by FA 900°C (**b**). The low and high current characteristics were recorded using an injection current density of 10^{-5} and 10^{-3} A cm^{-2}, respectively. Note the different scales for ΔV_{MG} and ΔV_{CC} (after [214])

accumulation of positive charges becomes significant, whereas the electron trapping in the bulk continues.

3.3.3 Annealing Dependence

Although the comparison in the previous chapter between MOSLEDs containing small and large REO_x clusters already contains an indirect annealing dependence, this chapter focuses on the annealing dependence mainly caused by hydrogen. Figure 3.18 displays the ΔV_{CC} characteristics recorded under an injection current density of 7×10^{-3} A cm^{-2} for Eu-implanted (solid lines) and unimplanted devices (dashed lines) for various annealing conditions. The unimplanted and untreated device exhibits a decrease of ΔV_{CC} up to a charge of $\sim 5 \times 10^{17}$ e$^-$ cm^{-2}, followed by

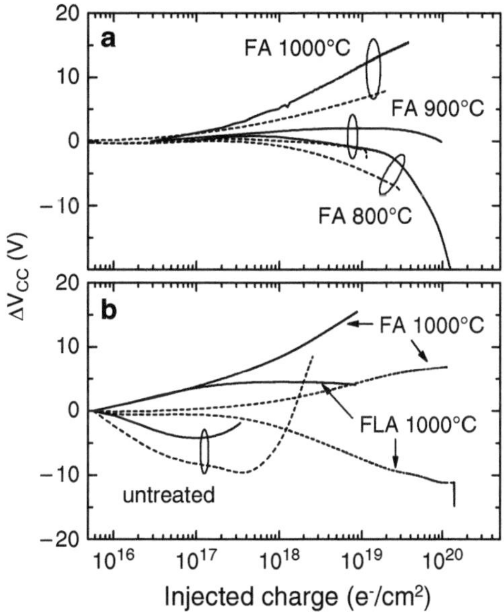

Fig. 3.18 Change of the applied voltage (ΔV_{CC}) under constant current injection for unimplanted (*dashed line*) and Eu-implanted devices with 1.5% Eu (*solid line*) as a function of the annealing temperature (**a**) and the annealing time (**b**) (after [132]). The injection current density is 7×10^{-3} A cm^{-2}

strong electron trapping and an early breakdown with increasing injected charges. The situation already changes by applying a low thermal budget (FLA 1,000°C or FA 800°C). In both cases, ΔV_{CC} stays constant up to $\sim 5\times10^{17}$ e$^-$ cm^{-2} followed by a more moderate decrease of ΔV_{CC}. At high thermal budgets (FA 1,000°C), only a slight but constant increase of ΔV_{CC} is observed which indicates a moderate electron trapping.

The behaviour of the unimplanted devices, especially that shown in Fig. 3.18b, can be strongly correlated with the hydrogen content given in Figs. 3.2–3.7. As it will be discussed in more detail in Sect. 3.4, the release of hydrogen is assumed to be jointly responsible for the creation and trapping of positive charges under charge injection. The Eu implantation just adds an electron trapping contribution to the unimplanted ΔV_{CC} characteristics. However, as shown in the previous chapter, the situation changes for high (Eu) and very high (Tb) thermal budgets which drastically modify the microstructure of the RE-implanted SiO_2 layers. The less intense electron trapping for Eu, FA 900°C in Fig. 3.18a compared to that in Fig. 3.17b is due to the different injection current densities of 7×10^{-3} A cm^{-2} and 10^{-3} A cm^{-2}, respectively.

3.3.4 *Concentration Dependence*

Figure 3.19 shows the ΔV_{CC} characteristics of devices implanted with different RE elements under an injection current density of 1.4×10^{-2} A cm^{-2}. The graph demonstrates that the introduction of additional electron traps is not limited to Eu but also

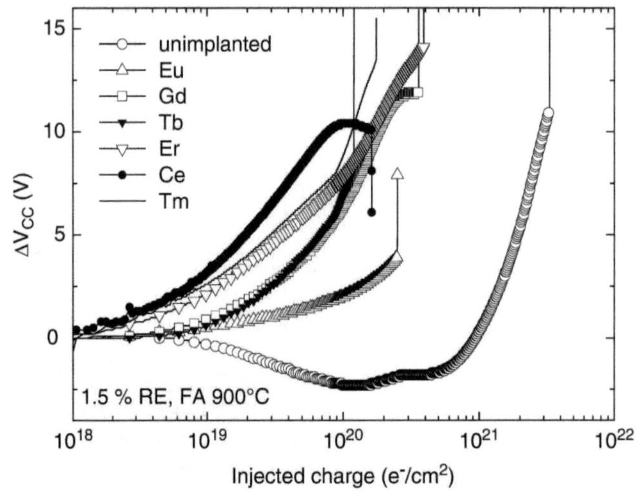

Fig. 3.19 Change of the applied voltage (ΔV_{CC}) under constant current injection for devices with different RE elements

applies to other RE elements. However, there are not enough data to decide whether a certain RE element introduces more traps than another. It should be borne in mind that Eu already forms large clusters in SiO_2 after FA 900°C, whereas the other RE-implanted MOSLEDs annealed at the same temperature are still characterized by small clusters.

In case of FA 900°C, the concentration dependence of different RE elements was measured, resulting in a similar set of characteristics for all investigated RE elements (Gd, Tb, Er, Ce, Tm) except Eu as shown in Fig. 3.20. Already the ΔV_{CC} characteristic of the device implanted with 0.25% Gd (Fig. 3.20a) differs considerably from that of unimplanted devices. Further increase of the implantation dose continuously shifts the ΔV_{CC} characteristic towards stronger electron trapping, but now on a smaller scale. In case of 0.5% Eu, the charge trapping is similar to that of other RE elements, but for the concentration range of 1.5–3% a reduction of the number of introduced electron traps is observed (Fig. 3.20b). The reason of this unusual behaviour is not yet known. As seen the concentration dependence is complex, but one conclusion can be drawn: the presence of RE impurities within the SiO_2 layer leads to additional electron trapping with a strong effect already at concentrations below 0.5% and a more moderate shift for higher concentrations.

3.4 The Charge Trapping and Defect Shell Model

This chapter tries to draw the overall picture regarding the electrical properties of RE-implanted MOSLEDs and is considerably based on results published in [214]. Generally, three classes of devices have to be distinguished: unimplanted devices,

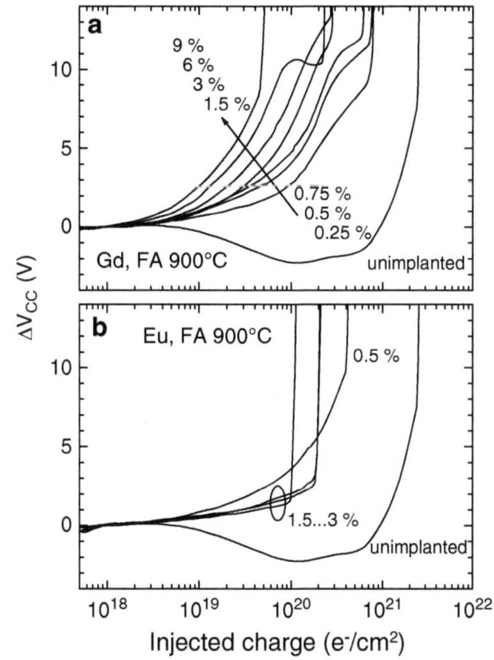

Fig. 3.20 Change of the applied voltage (ΔV_{CC}) under constant current injection for Gd-implanted (**a**) and Eu-implanted devices (**b**) for different RE concentrations

RE-containing devices with no or only small REO_x clusters and RE-containing devices with large REO_x clusters and significant RE agglomerations at the interfaces. In addition, the different phases of charge injection and the cases of low and high hydrogen content have to be considered. Therefore, the following discussion starts with the unstressed device, touches the different physical processes occurring under charge injection and introduces the defect shell model [214] which is used to explain the different features of charge trapping connected with the presence of REO_x clusters.

3.4.1 Before Charge Injection

The electrical properties of RE-implanted MOSLEDs are determined by two components: the behaviour of unimplanted devices and the effect of the RE implantation itself. The consideration starts with the thermally grown SiO_2 layer on Si which is regarded to be relatively free of charges. The subsequent deposition of the SiON layer introduces hydrogen which is assumed to be homogeneously distributed in the SiON layer with a drop down to lower values at the SiO_2–SiON interface (Fig. 2.7). The incorporation of hydrogen leads to the buildup of negative charges in the SiON layer and/or at the interface region to SiO_2 possibly not from the very beginning but in a very early phase of charge injection. However, it is believed that charges in the SiO_2–SiON interface region play the major role because of the following reasons: First, the effect of charges in this interface region on *IV*, *CV* and ΔV_{CC} characteristics is already reduced due to the long distance from the Si–SiO_2 interface, and charges within the SiON layer are even more affected by this; second, the SiON layer usually contains both electron and hole traps which results in an electrical compensation of the charges inside of the dielectric layer [223]; and third, the similarities of charge trapping in unimplanted Al–SiO_2–Si and ITO–SiON–SiO_2–Si structures imply that the charge trapping behaviour is mainly determined by the SiO_2 layer. However, the SiON layer, especially its interface region to SiO_2, can act as a significant source of positive charges during electron injection.

With annealing, hydrogen is driven out with increasing annealing temperature and time. In parallel, the negative charge caused by the incorporation of hydrogen (Sect. 3.2.1) decreases, which leads to a shift of the *IV* characteristic to lower voltages with increasing thermal budget (Fig. 3.10). At FA 1,000°C, the major part of hydrogen is driven out and the charges seen by the different characterization methods (*IV*, *CV* and ΔV_{CC}) are at minimum. The unimplanted device with FA 1,000°C is therefore assumed to be the most undisturbed device in the sense of an ideal MOS structure. As the negative charge is not seen by *CV* measurements, it might be balanced by an equivalent positive charging at the Si–SiO_2 interface.

In the as-implanted state, RE implantation leads to a shift of the *IV* characteristic to higher voltages (Fig. 3.11), indicating the buildup of an additional negative charge. According to Fig. 2.6, the energy deposition and thus the damage of the SiO_2 network is much higher around R_P than at the Si–SiO_2 interface. In addition, the shift of the *IV* characteristic is very similar for different RE elements, for which reason the additional negative charge is associated with implantation-induced defects and/or the trivalent nature of the RE ion and is assumed to have a charge centroid around R_P.

With annealing, three different processes take place:

1. Implantation-induced defects in SiO_2 will be partly annealed out
2. REO_x clusters form
3. RE ions, especially Eu, may diffuse toward the interfaces of the SiO_2 layer

Whereas the first process would imply a decrease of the negative charge introduced by implantation, the formation of REO_x clusters causes a considerable disturbance of the SiO_2 network which leads directly to the defect shell model developed by Nazarov et al. [214]. As RE elements have a larger atomic radius and a different valence electron configuration than Si, their incorporation into the SiO_2 network introduces defects or strained bonds close to the RE ions. The situation becomes even more severe in case of REO_x clusters which are possibly surrounded by a shell of defects and strained bonds. The extension of this shell increases with increasing cluster size, and thus, with increasing thermal budget. Whereas the defects may act as electron traps from the beginning, strained bonds are suitable precursors for the defect generation during charge injection. As there is a competition between Si and the RE elements for oxygen (Sect. 2.3.2), the defects are predominantly ODCs like the NOV. Whereas in case of small REO_x clusters the negative charge introduced by implantation (Figs. 3.11 and 3.13) is maintained, large clusters seem to have an affinity to positive charges which is both concluded from the results of *IV* and *CV* measurements (Figs. 3.12 and 3.14).

3.4.2 Physical Processes under High Electric Fields

With charge injection, a couple of different processes start whose intensities depend on the annealing and implantation conditions. Depending on how strong they are pronounced, the behaviour under constant current injection is different. These processes are:

1. Trapping of electrons by pre-existing traps. This applies to all electrons passing the SiO_2 layer and is less field dependent.
2. De-trapping of electrons. This may apply to devices starting with a high negative charge like untreated or FLA annealed devices and should be favoured by high electric fields (Sect. 3.3.1).
3. Anode hole injection [224, 225]. This is a process in which hot electrons reaching the SiO_2–anode interface can generate electron–hole pairs at the anode. Hot holes with an energy more than 4.5 eV (for a Si anode) are injected into the oxide and trapped at defects like Si–O strained and dangling bonds, ODCs and other structural defects of the amorphous SiO_2 network. In case of an ITO–SiON–SiO_2–Si structure, the ITO can be a source of holes, although the number of hot electrons in SiON which can reach the ITO layer is approximately two orders of magnitude lower than that in SiO_2 [204]. In addition, holes possibly injected from the ITO layer can already be trapped in the SiON layer, for which reason only anode hole injection from the SiO_2–SiON interface (with electron–hole pair generation in the SiON layer) can give a significant contribution.
4. Release of hydrogen being located at the SiO_2–SiON interface and, on a smaller scale, in the bulk of the SiO_2 layer by hot electrons. The released hydrogen diffuses as protons towards the injecting interface and starts to accumulate there. If there are negative charges around R_P, the diffusing protons may also cause a partial charge recombination there. The release of hydrogen depends on the kinetic energy of the impinging electrons and thus on the local electric field. The hydrogen release can also be connected with the generation of negative charges. It was shown that the generation of electron traps with generation cross sections in the order of 10^{-18}–10^{-20} cm^2 is directly associated with a hydrogen release according to

$$X - H + e^{**} \rightarrow X^{-} + H^{+} + e^{*}, \tag{3.10}$$

with X and $e^{**/*}$ being the defect bonded to hydrogen and the exciting electron before and after its interaction with the defect, respectively [226]. However, because of the small cross section, this specific process is relevant for injected charges above 10^{18} e$^-$ cm^{-2} only.

5. Trap and defect generation by band-to-trap ionization for high-energetic electrons. In this case, a charge is generated by the impact of hot electrons on filled traps, weak bonds or other defect precursors. In case of an asymmetric distribution of electron traps in the SiO_2, an asymmetric positive charge generation regarding the positive and negative polarity of the applied voltage to the gate electrode is observed. It was suggested [227] that the NOV is one of the defects that can emit an electron by impact ionization.
6. Trap and defect generation by band-to-band ionization in case of high-energetic electrons. This process permanently occurs but dominates for high-level charge injection and is assumed to be responsible for oxide degradation and the final breakdown. Although the defect generation is assumed to increase with increasing electric field, the *charge*-to-breakdown value is independent of the injection current in a first approximation (which leads to the well-known reciprocal relation between injection current and *time*-to-breakdown value). Therefore, the field dependence of trap and defect generation is assumed to be weak in the observed range of injection current densities. As the energy distribution of the hot electrons also depends on the location within the SiO_2 layer (Sect. 3.1.2), the favourite location for this process is in the bulk of the SiO_2 layer but close to the SiON interface.

The processes are schematically shown in Fig. 3.21.

3.4.3 The Different Phases of Charge Injection

The first *phase of low-level charge injection* (10^{14}–$10^{16}\,e^-\,cm^{-2}$) is characterized by trapping/de-trapping in pre-existing traps and the generation of positive charges. In case of unimplanted devices, this is characterized by the buildup of a positive net charge in the SiO_2 layer (Fig. 3.16a) which is the stronger the higher the injection current (Fig. 3.15a) and the higher the hydrogen content (Fig. 3.18) are.

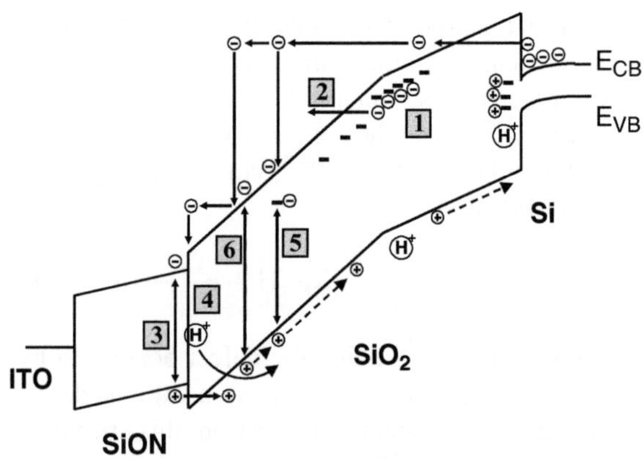

Fig. 3.21 Processes of charge generation and trapping in an ITO–SiON–SiO_2–Si structure: trapping in pre-existing traps (1), de-trapping (2), anode hole injection from SiON (3), hydrogen release (4), band-to-trap ionization (5), band-to-band ionization (6)

As unimplanted devices with a high hydrogen content start with a larger negative charge, the buildup of a positive net charge under charge injection is equivalent to the reduction of the initial negative charge, most probably by de-trapping (process 2). At the same time, the negative shift of the *CV* characteristics (Fig. 3.15d) points to the buildup of a real positive charge at the Si–SiO_2 interface. Positive charges are continuously generated by the processes 3–6 and migrate towards the Si–SiO_2 interface under the applied electric field. As the hydrogen release is often associated with low capture cross sections [226], this process may become significant for medium- and high-level charge injection only. In case of band-to-trap ionization the NOV is assumed to be the most relevant defect.

RE implantation creates defects either directly by the energy deposition during implantation or indirectly in combination with annealing by the formation of REO_x clusters which are surrounded by a defect shell. Some of these defects are assumed to have a very large electron capture cross section in the order of $10^{-14}\,cm^2$ for which reason an injected charge as low as $10^{14}\,e^-\,cm^{-2}$ can already induce a strong negative charge. A similar trap was generated by hot electron injection into the SiO_2 layer of a MOS transistor and was attributed to Si dangling bonds in the SiO_2 [228]. The large cross section would explain the shift of the "initial" *IV* characteristics in Fig. 3.11 (small but non-zero charge injection), the first slope of the ΔV_{CC} characteristics in Fig. 3.16b and the *CV* behaviour in Fig. 3.15e. Generally, the ΔV_{CC} characteristics of RE-implanted structures *always* exhibit a stronger electron trapping (or a smaller positive charge accumulation) than the unimplanted counterparts throughout the *whole* range of charge injection! This additional RE-induced trapping superimposes with the aforementioned processes in unimplanted structures, which is why the accumulation of positive charges is also seen in the corresponding ΔV_{CC} and ΔV_{MG} characteristics (Fig. 3.16b).

Whereas RE-implanted structures with small REO_x clusters show a strong electron trapping in the bulk at the very beginning followed by a smooth decrease, structures with large REO_x clusters exhibit strong electron trapping throughout the whole range of charge injection (Fig. 3.17b). This is caused by the generation of additional electron traps which can be correlated with the higher degree of disorder or structural damage of the SiO_2 matrix around large REO_x clusters and thus a larger extension of the defect shell. In addition, a positive charge accumulation at the Si–SiO_2 interface is not detectable. Due to the low hydrogen content, the release of hydrogen is low, and the positive charge generation caused by anode hole injection from SiON might be outbalanced by the strong electron trapping. Although the cluster sizes in cases of Tb 1,100°C and Eu 900°C are comparable, the low-level charge trapping behaviour of Tb 1,100°C is between that of Tb 900°C and that of Eu 900°C. Therefore, it is concluded that the EuO_x–SiO_2 interface region contains more defects than the TbO_x–SiO_2 interface region.

The *phase of medium-level charge injection* (10^{16}–$10^{18}\,e^-\,cm^{-2}$) is a transition phase in which the influence of pre-existing traps decreases and in which the creation of new defects through hot electrons becomes more significant. The buildup of a positive net charge in the bulk reaches a maximum and turns into electron trapping, whereas the accumulation of positive charges at the Si–SiO_2 interface continues. This applies for unimplanted structures and devices with small REO_x clusters, whereas the strong electron trapping in devices with large REO_x clusters continues. However, the onset and the intensity of electron trapping differ depending on the hydrogen content and the injection current density which e.g. explains the partly different results for unimplanted devices in Figs. 3.16a and 3.18. For application purposes, it is worth noting that the Tb-implanted device annealed at FA 900°C does not show charge trapping in the bulk for more than two orders of the injected charge (Fig. 3.16b). For Tb, FA 1,100°C, the charge trapping in the range of medium-level charge injection can be described by traps with capture cross sections of 2×10^{-17} and $5 \times 10^{-18}\,cm^2$ which are associated in literature with the Si–O• defect [229] or ODCs [230], and with "water-related" traps possessing the configuration of ≡ Si–OH [231]. As already mentioned, the generation of additional electron traps for devices with larger REO_x clusters can be related to a larger disorder of the SiO_2 network and thus larger defect shells around these clusters.

The last *phase of high-level charge injection* ($>10^{18}\,e^-\,cm^{-2}$) is dominated by the creation of new defects, which is usually connected with an enhanced electron trapping in the bulk and positive charge accumulation at the Si–SiO_2 interface. This behaviour can be seen in Figs. 3.16, 3.17, 3.19 and 3.20. The creation of new defects is still accompanied by the processes of anode hole injection and hydrogen release. In some cases, at very high levels of the injected charge ($>10^{20}\,e^-\,cm^{-2}$), just before the electrical BD of the dielectric occurs, a positive charge trapping in the bulk of the dielectric is observed. This effect can be attributed to the redistribution or extension of captured positive charges from the SiO_2–Si interface towards the SiO_2 bulk beyond the tunnelling distance as these positive charges become now visible in ΔV_{CC} measurements. Generally, this effect is a strong indication for the continuing degradation of the SiO_2 matrix (Chap. 6).

Table 3.1 summarizes the behaviour of such structures under constant current injection.

Table 3.1 Charge trapping processes in unimplanted and RE-implanted MOSLEDs

Phase of charge injection	10^{14}–$10^{16}\,e^-\,cm^{-2}$	10^{16}–$10^{18}\,e^-\,cm^{-2}$	$>10^{18}\,e^-\,cm^{-2}$
General processes	Trapping/de-trapping in pre-existing traps	Transition between the traps of pre-existing traps and that of new created traps	Defect and trap creation
	PCG	PCG	PCG
Unimplanted	Reduction of initial neg. charges in the bulk	Weak NCT or PCA in the bulk	Strong NCT in the bulk, but less than RE implanted
	PCA at the IF	PCA at the IF	PCA at the IF
Small REO_x clusters	Strong NCT in the bulk	Weak or no NCT in the bulk (plateau)	Strong NCT in the bulk
	PCA at the IF	PCA at the IF	PCA at the IF
Large REO_x clusters	Strong NCT in the bulk	Strong NCT in the bulk	Strong NCT in the bulk PCA at the IF

Abbreviations: *PCG* positive charge generation, *PCA* positive charge accumulation, *NCT* negative charge trapping, *IF* Si–SiO_2 interface

Chapter 4
Electroluminescence Spectra

The electroluminescence spectra obtained from RE-implanted MOSLEDs are basically a result of an electronic transition within the $4f$ shell of a trivalent RE ion after an electrical excitation by hot electrons. The optical spectra of REs, mostly obtained by PL or absorption measurements, have been widely investigated for decades, especially in crystalline host materials like $LaCl_3$ or Y_2O_3, and there are a couple of excellent surveys about this topic [233–238]. Therefore, only a brief introduction concerning the physics of $4f$ intrashell transitions is given in Sect. 4.1.1, as these explanations can be found in greater detail in most of the textbooks devoted to quantum mechanics and the literature mentioned above. In the following, the EL properties of RE-implanted MOSLEDs are shown and discussed on the basis of own results. While the Sects. 4.1.2–4.1.4 deal with the used experimental techniques, with the EL spectra of unimplanted and with that of RE-implanted MOSLEDs, the subsequent chapters cover the dependence of the EL spectrum on the operation and fabrication conditions. In detail, the focus is set on the injection current dependence (Sect. 4.2), the concentration quenching (Sect. 4.3) and the annealing behaviour (Sect. 4.4).

4.1 Spectral Features

4.1.1 4f Intrashell Transitions in Trivalent Rare Earth Ions

In general, the electronic structure of a neutral RE atom is composed of a Xe electronic configuration, a partly filled $4f$ shell, sometimes an electron in a $5d$ orbital (the row of lanthanides shows some irregularities on this aspect) and two electrons in the $6s$ shell [239]. The $6s$ and the $5d$ electrons are loosely bound, for which reason RE elements try to achieve a trivalent ionic configuration and to form oxides preferentially as RE_2O_3 (Sect. 2.3.2). Consequently, the electronic configuration of RE^{3+} ions can be written as $[Xe]4f^n$ with n being the number of electrons in the $4f$ shell, which increases steadily from zero for the element La up to 14 for Lu. The interaction of the electrons among each other within the $4f$ shell can be described by the Russel–Sanders coupling where the orbital angular momentums l_i of the electrons have to be added as vectors giving the total orbital angular momentum L. In the same way, the spins of the electrons sum up to the total spin S, and because of the spin–orbit coupling, L and S can be added to a total angular momentum J. The different possibilities to perform a vector addition define the different electronic arrangements within the $4f$ shell and thus the energy levels of the RE^{3+} ion, which are labelled with the well-known notation $^{2S+1}L_J$. Because of the Pauli exclusion principle, the total wavefunction must be antisymmetric, which excludes half of the L values for a given S value.

The energy levels are grouped in accordance with Hund's rules where the L and J values define the main energy levels and the sub-levels, respectively. In case of crystalline host materials, the interaction with the local electric field of the matrix causes a further splitting of the J levels, which is called the crystal field splitting. In non-crystalline media, the variation of the local chemical environment leads to a broadening of the crystal field levels or to the formation of an energy band whose width is in the same order than the crystal field splitting. Although the prediction of energy levels in such systems is complex and requires advanced quantum mechanical tools [234], as a very rough rule of thumb, one can state that the energy levels of different L are separated by energies in the order of a few hundreds of meV up to a few eV and that energy levels of different J are separated by energies in the order of a few tens up to a few hundreds of meV. Crystal field splitting normally accounts for separation energies in the order of a few tens of meV and below. The splitting of the energy levels according to these rules is exemplarily displayed for two electrons with $l_i = 3$ in Fig. 4.1. The comparison with the real world of Pr^{3+} shows that the lower lying levels are reproduced quite reasonably, but that there are larger differences regarding the higher lying levels.

In contrast to $5s$ and $5p$ electrons, the electrons of the $4f$ shell are closer to the nucleus, for which reason they are well screened from the local chemical environment. In fact, the interaction strength of the anion ligands with the $4f$ electrons in RE-doped halogenides and chalcogenides is by a factor of 20–50 weaker than with the $5d$ electrons [236]. As a result, the characteristic emission lines from $4f$ intra-shell transitions are narrow and relatively fixed in position. In amorphous materials such as thermally grown SiO_2, the FWHM is broader than in crystalline hosts and reaches values up to 100 meV in the present case of EL.

$4f$ intra-shell transitions are optical dipole forbidden transitions resulting in a low oscillator strength. Thus, the efficiency in optical absorption is low. However, whether the radiative deexcitation of an excited state is efficient or not depends on the presence of competing non-radiative deexcitation channels, and may differ from case to case. Nevertheless, the radiative decay constant of such an excited state is expected to be long, and indeed, decay times for the main EL emission lines of the different RE^{3+} ions in the order of a few 100 μs up to a few ms have been measured for RE-implanted MOSLEDs. To obtain an efficient luminescence, the restrictions of optical selection rules have to be circumvented, which can be done in an optical or non-optical way. The efficient optical excitation would require the lifting of an electron into the $5d$ shell (which is not dipole forbidden) followed

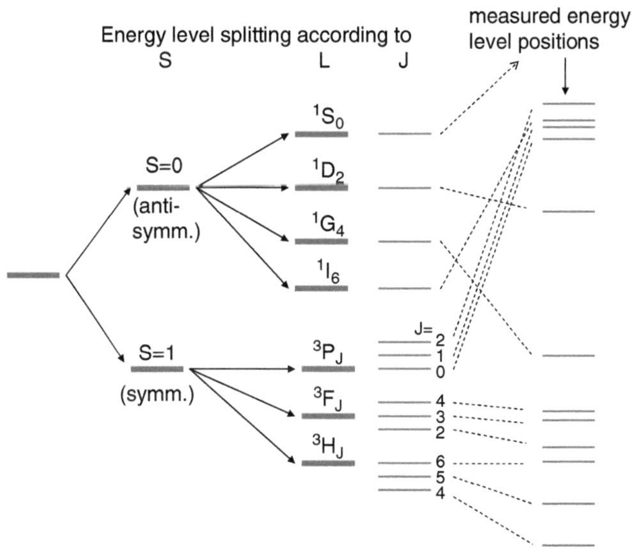

Fig. 4.1 Basic scheme of the energy level splitting of a two-electron system due to spin–orbit coupling and the comparison with the measured positions in case of Pr^{3+}. The latter ones are given in [233]

by a radiative transition of the system back to the ground state or followed by a transition – regardless if radiative or not – to an excited $4f$ configuration. Typical excitation and emission wavelengths for transitions involving a *5d* orbital lie in the UV and blue-green spectral region. Non-optical methods for an efficient excitation of RE^{3+} ions can take advantage of energy transfer processes from sensitizers such as Si or Ge NCs (Sect. 5.3) or other RE elements (Sect. 5.4), or use hot electrons for impact excitation. To avoid confusion, it should be noted that energy transfer processes are counted among non-optical methods independent of the way the energy to be transferred is generated. So the energy transfer from Si–NCs to Er^{3+} ions described in literature (Sect. 5.3.1) is regarded to be non-optical in nature even if the Si–NCs are optically excited. For the sake of completeness, another energy transfer process should be mentioned, namely the excitation of RE^{3+} ions by the energy released in charge carrier recombination as it happens in case of RE ions incorporated in GaAs and related III–V materials [240]. Of course, the boundaries within this classification are floating; for example the electronic transition of an electron to the ground state can be understood as the recombination of an electron-hole pair in the broadest sense.

If one compares PL and EL, the following differences are to be noted. Usually monochromatic light produced either by a laser or by a lamp in combination with a monochromator is used to excite the PL. The excitation wavelength can fit an absorption wavelength of the RE^{3+} ion or not, which is called a resonant or non-resonant excitation, respectively. Clearly, the resonant excitation is expected to be much more efficient than a non-resonant excitation. Sample preparation is easier compared with the case of EL, and because of the possibility for a selective excitation, PL is frequently used for investigations devoted to the physical background of a system. In contrast to this, EL – as far as it is based on *impact excitation* of hot electrons – is a broadband excitation as the energy distribution of these hot electrons is very broad (Sect. 3.1.2). In addition, the optical selection rules do not apply, for which reason the EL has the potential to reach a higher efficiency than the PL (For questions how to exploit this potential and for the different efficiency definitions, see Chap. 5). The sample preparation for EL is more complex and costly, but also much more close to a real application.

4.1.2 Electroluminescence and Decay Time Measurement Techniques

This chapter presents a brief overview of the methods used to obtain the EL spectra and EL decay times presented in *this* book. For recording the EL spectra and for monitoring the EL intensity as a function of time or injection current, a source measurement unit was used to apply either a constant current (EL spectra and EL intensity over time) or a voltage ramp to the device. Usually, the device is placed on a chuck mounted on a microscope table and is electrically contacted by the needle of a probe and the conducting surface of the chuck. The light is collected by a suitable objective and guided by a glass fibre to a monochromator in order to maximize the collection efficiency. The light signal is then detected by a photomultiplier, a CCD (charge-coupled device) camera or an InGaAs detector for the IR spectral region. The CCD camera allows the recording of kinetic series, i.e. the evolution of the EL spectrum with time. As the glass fibre is opaque in the UV, EL measurements in this spectral region were performed by placing the device in front of the entrance slit of the monochromator. While the measurement of the EL power efficiency is explained in Sect. 5.1.2 in more detail, the experimental setup for all other measurements is sketched in Fig. 4.2. If not otherwise stated, all measurements were performed at RT.

All spectra recording techniques that were used in the course of these investigations are wavelength-selective techniques, for which reason all spectra in the subsequent chapters are given on a wavelength scale. However, distances in an energy level system and the width of spectral lines are usually given in energy units, most of all in eV. The energy E_{Ph} of a photon is related to its wavelength λ by the equation $E_{\mathrm{Ph}} = hc/\lambda$, allowing the simple transformation of the wavelength scale into an energy scale and reverse. Doing so, it should be considered that such a transformation is not an equal-area projection; that is transforming the wavelength into an energy scale without changing the shape of the spectrum will shift the meaning of the integral: while the integral over an EL line on the wavelength scale gives a number proportional to the number of emitted photons, the integral over an energy scale is proportional to the emitted energy at this wavelength.

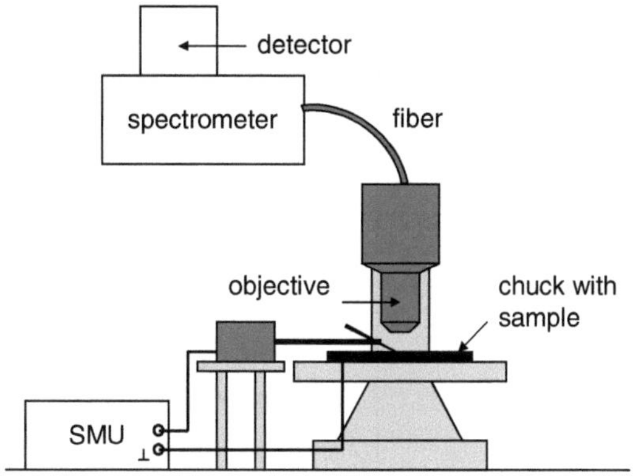

Fig. 4.2 Typical setup for EL measurements

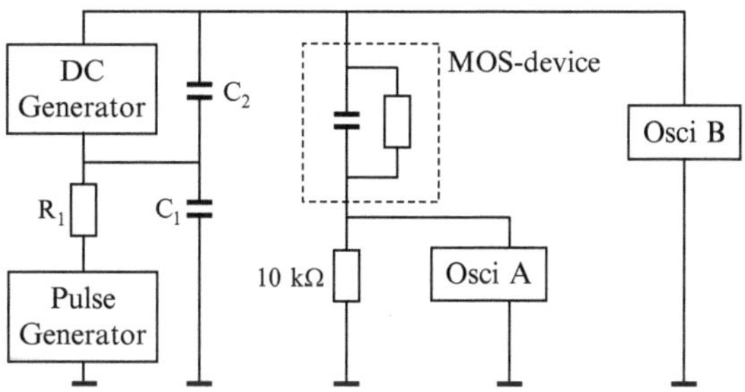

Fig. 4.3 Electric circuitry used for EL decay time measurement. The oscilloscope channels A and B record the pulse shape of the current and the applied voltage, respectively. Note that the DC generator must allow floating ground operation

For EL decay time measurements, a pulsed voltage has to be applied to the device, whereby the time constant of the pulse edges should be significantly shorter than the time constants to be measured. In the best case, a suitable voltage pulser providing the right voltage levels and the required time resolution is available. There is a large variety of pulsers and function generators for low voltages, whereas in case of high voltages suitable devices are less well available and expensive. An alternative is to create the pulsed voltage by a circuitry adding the output of a low voltage pulser to the constant voltage of a second device. In this case, the pulsed voltage does not switch between zero and the operation voltage, but between a low and high voltage point on the IV characteristic. The low voltage should favourably lie on a level where FN tunnelling has not yet been started to ensure that there is no EL in the interval between two pulses. The circuitry used first in [100] is shown in Fig. 4.3, with the resistance R_1 and the capacitance C_1 determining the time constant of the voltage pulse and C_2 bridging the constant voltage generator for AC pulses. Note that the constant voltage generator must allow floating ground operation! This method gives a quick access to EL decay time measurements without expensive high voltage pulsers, but undoubtedly, it is less comfortable, and the voltage pulse level difference is restricted to the voltage span of the low voltage pulse generator.

A simple method to record the time-resolved EL signal is the use of a common Si photodiode in combination with an oscilloscope. This method is applicable if the EL

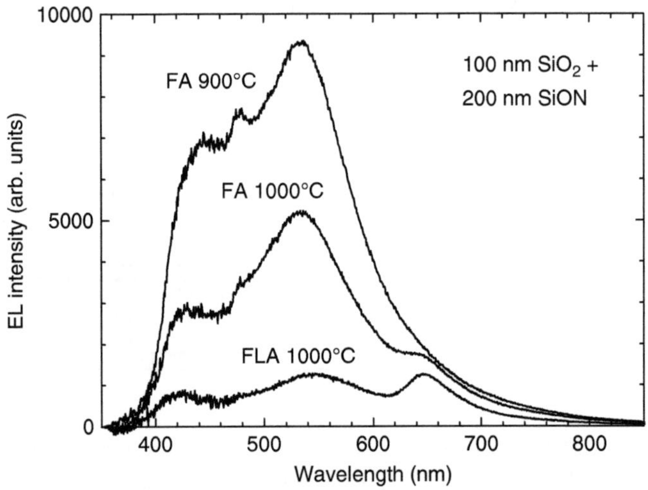

Fig. 4.4 EL spectra of unimplanted MOSLEDs after different thermal treatments. All spectra were recorded at a constant injection current density of 1.4×10^{-3} A cm^{-2}

signal is strong, but for weaker signals, more advanced detection techniques have to be used. In our case, the signal of the detector was amplified by a fast 300 MHz amplifier and recorded by a multichannel scaler.

4.1.3 Electroluminescence of Unimplanted MOS Structures

Similar to the electrical properties, the EL spectrum of unimplanted SiO_2 layers depends on the post-oxidation anneals and is influenced by the presence of the SiON layer. Figure 4.4 shows such spectra from a device with a 100 nm thick SiO_2 layer, a 200 nm thick SiON layer and a top electrode made of ITO. The spectra are characterized by a broad emission band in the visible with maxima around 420–450 nm, 530–550 nm and 650 nm. While the EL intensity is low for low and high thermal budgets, larger EL intensities are achieved for medium thermal budgets. This optimum might be the result of two processes running in opposite directions with increasing temperature: the annealing out of LCs and the annealing out of defects that act as luminescence quenchers. In addition, one type of defect can be transformed into another one during annealing. However, for 100 nm thick SiON layers, the red part of the EL spectrum is more pronounced than the green one, and in some cases, the blue spectral region is dominating. Nevertheless, in all investigated cases the measured power efficiency was below 10^{-6}.

Most of the defects discussed in Sect. 2.2 can serve as potential luminescence centres. Frequently, luminescence bands around 460 nm, 560 nm and 660 nm are found in PL and/or EL investigations of stoichiometric and Si-rich SiO_2, which are assigned to the NOV [167, 241, 242] or Si_2^0 centre [28], to one of the different E' centres [243, 244] or to the NBOHC [74, 245], respectively. Sometimes, a PL band at 415 nm is observed and attributed to a weak oxygen bond in interstitial oxygen [246, 247]. For more details, it is referred to the literature [67, 163, 247]. As SiON has normally much more defects than thermally grown SiO_2, these defects can give a significant, if not even a major contribution to the observed EL spectrum. In [248], three EL bands peaking around 400, 520 and 690 nm were observed in amorphous silicon nitride and all assigned to radiative recombinations between electrons and holes in the tail states of the conduction and valance bands with the involvement of different nitrogen defects and Si dangling bonds. Green EL bands originating from SiN_x layers between 530 and 560 nm were also found in [172, 173, 249–252], whereby in the latter case the green band was produced by a post deposition annealing step in an oxidizing atmosphere. Orange–red EL was found in [253] and [254, 255] with EQE around 2×10^{-5} and 1.2×10^{-5}, respectively. Furthermore, NOV defects, which are also present in the SiON layer at lower concentrations (Sect. 2.2.2), can give an additional contribution in the blue spectral region. Interestingly, in many cases where EL from SiN_x or SiO_xN_y films is reported, the device structure is not basing on a Si wafer but on a glass substrate covered by an ITO layer.

In summary, unimplanted MOSLEDs exhibit broad but weak EL in the visible with a power efficiency below 10^{-6} (for definition see Sect. 5.1.1), which is most probably composed of a blue component due to ODCs, a green component originating from structural defects in SiON and a red component originating from NBOHCs. The contribution from any kind of Si clusters is unlikely as the slight local Si excess that might occur due to stoichiometric fluctuations is enough for the formation of ODCs, but too small for Si nanocluster formation. Possibly the major part of the EL originates from the region around the SiO_2–SiON interface because of the lower defect density in SiO_2 and the more unfavourable excitation conditions for EL in the bulk of the SiON (Sect. 5.2.3).

4.1.4 Electroluminescence Spectra of Rare Earth Ions in Silicon Dioxide

In the following, we present the EL spectra of the RE elements that were used in our RE-implanted MOSLEDs [129, 256]. These are, in order of their appearance in the periodic system of elements, Ce, Pr, Eu, Gd, Tb, Er, Tm and Yb. In principle, all RE elements should be electrically excitable in a MOS structure, but this happens with quite different efficiency. In most of the EL spectra, the line positions of relevant $4f$ intra-shell transitions are given according to [233]. In these cases, the energy level positions of RE^{3+} ions in hexagonal crystals (mostly $LaCl_3$) were used, and only the average energy over different crystal field levels belonging to the same (L, S, J) triple was considered. Although amorphous SiO_2 is a quite different matrix, the observed EL line positions fit quite well with those of crystalline hosts due to the screened nature of the $4f$ shell.

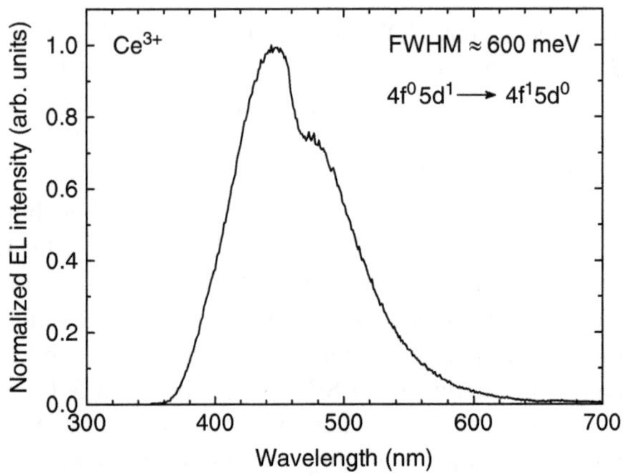

Fig. 4.5 Typical EL spectrum of Ce-implanted MOSLEDs

4.1.4.1 Cerium

Ce^{3+} has only one electron in the $4f$ shell, for which reason there is only one ground state with $L = 3$ and $S = 1/2$ that splits into the levels $^2F_{5/2}$ and $^2F_{7/2}$ with a separation energy of about 280 meV [233]. Any excitation with an energy of a few eV can lift this electron at least into the $5d$ shell, and the radiative transition of this electron back to the ground state causes the blue EL band around 440 nm shown in Fig. 4.5. As the $5d$ shell is sensitive to the local environment and because this environment varies considerably in the amorphous SiO_2 network, the EL band is broad giving a FWHM in the order of 600 meV. This transition is not dipole forbidden, and indeed a short EL decay time in the order of 75 ns was measured [130]. As the absorption wavelengths of Ce^{3+} overlap with the emission wavelength of Gd, the latter can be used to pump Ce (Sect. 5.4.1).

4.1.4.2 Praseodymium

With two electrons in the $4f$ shell, the energy level system of Pr^{3+} has a moderate complexity yet with a $^3H_4$ ground state (Fig. 4.1), but the different combinations of the existing levels result in a large number of possible transitions. Figure 4.6 shows the EL spectrum of a Pr-implanted SiO_2 layer annealed at FA 900°C, which was recorded at a constant injection current density of 0.14 A cm^{-2}. Possible transitions from the ^3P_J levels to $^3H_J (J = 4, 5, 6)$ and ^3F_J levels $(J = 2, 3, 4)$ are marked in the graph, indicating that the assignment of single EL lines to a specific transition is difficult. Moreover, transitions from the $^1I_6$ level were not considered in Fig. 4.6, although possible, as its energetic position is close to that of the $^3P_1$ level. However, the Pr-implanted devices exhibited such a low EL power efficiency ($<10^{-6}$) that they were not considered for further investigations.

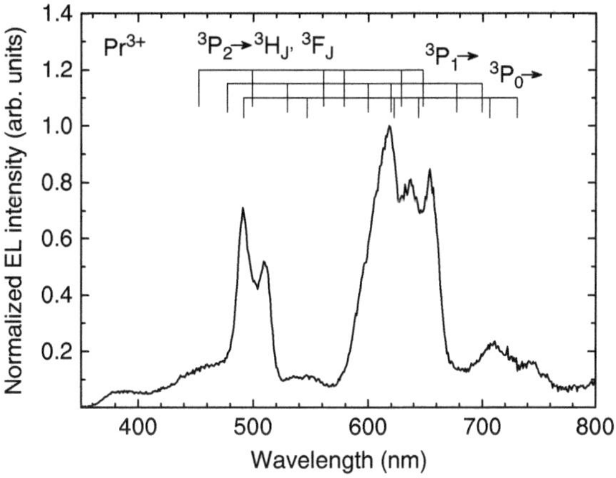

Fig. 4.6 Typical EL spectrum of Pr-implanted MOSLEDs. Three groups of possible transitions from the $^3P_2$, $^3P_1$ and $^3P_0$ level to ^3H_J ($J = 4, 5, 6$) and ^3F_J levels ($J = 2, 3, 4$) are given

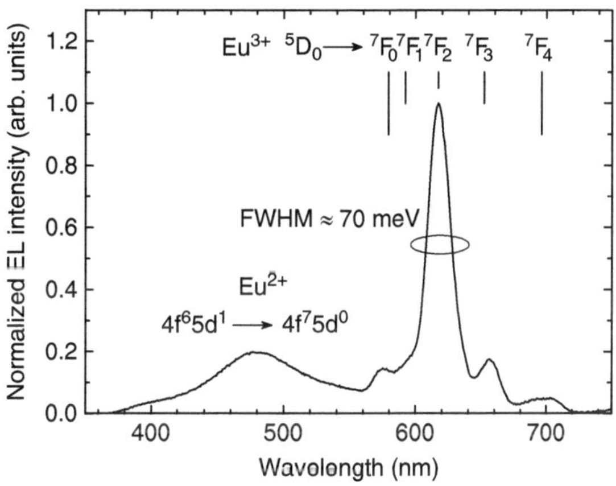

Fig. 4.7 Typical EL spectrum of Eu-implanted MOSLEDs

4.1.4.3 Europium

The next element in the course of this study is Eu whose spectrum given in Fig. 4.7 also shows some specialities. Within the row of RE elements, Eu features a maximum in the ionization energy from $Eu^{2+} \rightarrow Eu^{3+}$ [239], with the practical consequence that Eu^{2+} and Eu^{3+} may coexist in the SiO_2 matrix. Hence the EL spectrum is composed of a couple of narrow lines in the red that are characteristic for the emission of Eu^{3+} and a broad EL band in the blue-green spectral region that is believed to be mainly due to the emission of Eu^{2+} ions. Eu^{3+} has 6 electrons in the $4f$ shell, producing a total spin of 3 (all spins parallel) and a multiplicity of 7 for the ground state. Thus, the observed red EL lines originate from a transition of the $^5D_0$ state to the ground state sub-levels ^7F_J with $0 \leq J \leq 6$ and with $^5D_0 \rightarrow {}^7F_2$ being the most intense transition. This main line around 618 nm is characterized by a FWHM of about 70 meV and an EL decay time between 350 and 700 μs [190] depending on the specific device preparation conditions.

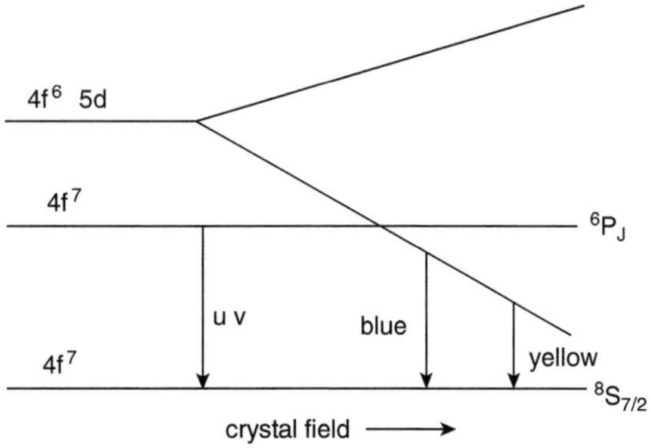

Fig. 4.8 Schematic energy level diagram of the Eu^{2+} ion as a function of the magnitude of the crystal field according to [235, 268]

For the emission caused by Eu^{2+}, a similar argumentation holds as in the case of Ce^{3+}. In the ground state $^8S_{7/2}$, the $4f$ shell of Eu^{2+} is half filled with 7 electrons whose spins are all parallel aligned. As schematically shown in Fig. 4.8, the energetic position of the $4f^65d^1$ configuration strongly varies with the crystal field. As the next excited state within the $4f$ shell, ^6P_J, lies at a higher energetic position in most cases, the electric excitation will lift most probably one electron from a $4f$ orbital into a $5d$ one. The radiative deexcitation to the ground state is not dipole forbidden, resulting in a short EL decay time constant, and the involvement of the weakly screened $5d$ orbital causes a broad FWHM. The shape and relative intensity of this blue EL band are very sensitive to the preparation and operation conditions of the MOSLED, for which reason the variability of the Eu spectrum is a topic for subsequent chapters. Although the EL intensity in the blue is higher than in the case of unimplanted MOSLEDs, the blue EL band may contain minor contributions originating from the SiON layer, its interface or from ODCs in the SiO_2 layer.

4.1.4.4 Gadolinium

The ground state $^8S_{7/2}$ of the trivalent ion of the following element, Gd^{3+}, has the same electronic configuration as Eu^{2+} with 7 spin-parallel electrons. As $L = 0$, there is no splitting of the ground state to different J sub-levels. In contrast to Eu^{2+}, the energy required to lift an electron from the $4f$ into the $5d$ shell is higher than the energetic difference to the next $4f$ multiplet ^6P_J with 6 spin-parallel electrons and one electron having an anti-parallel aligned spin. Nevertheless, the energies of the $^6P_J \rightarrow {}^8S_{7/2}$ transitions are large, causing a sharp EL line around 316 nm and weaker satellites at 310 and 325 nm (Fig. 4.9). In the order of increasing wavelength, the EL lines are assigned to a transition from the ^6P_J state with $J = 3/2, 5/2, 7/2$ to the ground state (see Fig. 4.9), but in comparison with the transition energies given in [233], they are shifted by roughly 10 nm to the red, which is equivalent to an energy difference of about 130 meV. This might be due to the fact that in this case the local electric field of the amorphous SiO_2 network differs

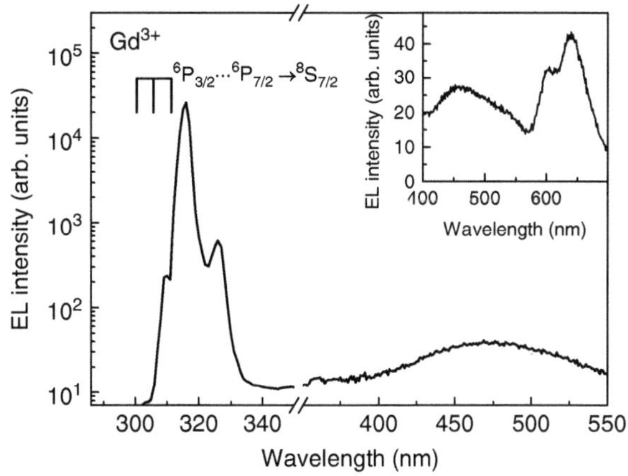

Fig. 4.9 Typical EL spectrum of Gd-implanted MOSLEDs. The inset shows an EL spectrum where the UV emission of Gd^{3+} was suppressed by an edge filter (after [127])

significantly from that of a crystalline host, and that the values given in [233] were obtained at a temperature of 1.5 K. The peaks are characterized by a FWHM of less than 40 meV, and EL decay times between 1.2 and 1.8 ms were measured for the 316 nm line [135]. A more detailed investigation about the energy level system of Gd^{3+} can also be found in [257, 258].

In addition, a broad but weak EL band extending from the blue up to the red can be observed whose integral EL intensity is two orders of magnitude lower than that of the 316 nm line (Note the logarithmic scale of the EL intensity). The inset of Fig. 4.9 shows an EL spectrum with two main EL bands in the visible range, which are assumed to originate from ODC and NBOH centres as typical network defects of the amorphous silicon dioxide created by ion implantation (Sect. 4.1.3). The bands might be excited either directly by inelastic scattering of hot electrons or by an energy transfer process from excited Gd^{3+} ions. As the EL decay times of Gd^{3+} are rather long, an energy transfer process from excited Gd^{3+} ions – which would shorten the EL decay time – is less probable or occurs only with minor intensity. Therefore, the mechanism of electron impact excitation is the most probable mechanism, as it is also responsible for the weak defect EL observed from unimplanted SiON–SiO_2 structures.

4.1.4.5 Terbium

Moving on in the periodic system of elements, Tb^{3+} with 8 electrons in the $4f$ shell is the next RE of interest. The energy level system offers two excited states, the $^5D_4$ and the $^5D_3$ one, from which a couple of transitions to the ground state ^7F_J are possible. As seen in Fig. 4.10, most of the lines of these two groups of transitions are visible in the EL spectrum. The main emission line $^5D_4 \rightarrow {}^7F_5$ around 545 nm has a FWHM of about 50 meV and is characterized by a bimodal shape, in which the energetic difference between the two sub-peaks of less than 30 meV is in the same order as the crystal field splitting of the $^7F_5$ state (that of $^5D_4$ is considerably

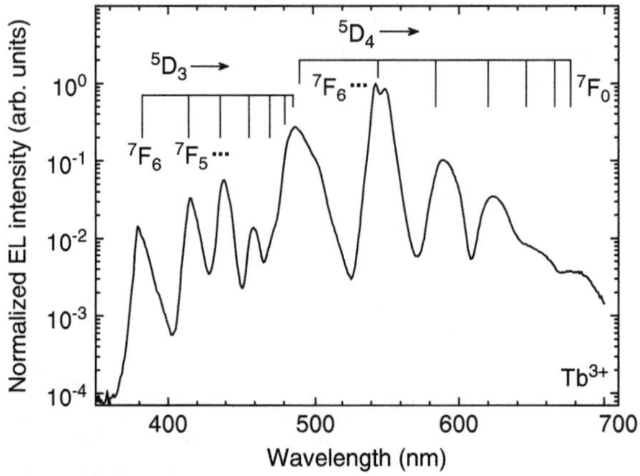

Fig. 4.10 Typical EL spectrum of Tb-implanted MOSLEDs

less). The EL intensity scale of Fig. 4.10 is a logarithmic one, so the $^5D_4 \rightarrow {}^7F_5$ line normally covers more than 50% of the overall emission (57% in the case shown in Fig. 4.10). Compared to other RE elements, the Tb emission with an EQE of 16% is the most efficient one, and the main emission wavelength is close to 555 nm where the photopic luminosity function (human eye sensitivity at day light) has its maximum [259]. Moreover, the existence of two groups of emission lines and the shift of their EL intensity ratio will give some additional insights into the EL excitation and quenching mechanism (Sect. 4.3.2).

4.1.4.6 Erbium

Er is the most investigated RE element in the field of Si based light emission. Its trivalent ion Er^{3+} has 11 electrons in the $4f$ shell, and the most interesting transition lies within the ground state multiplet, namely the $^4I_{13/2} \rightarrow {}^4I_{15/2}$ transition emitting at about 1,540 nm. However, there are a couple of higher lying states giving rise to additional EL lines in the visible spectral region. Figure 4.11 shows the EL spectrum of a MOSLED implanted with 0.5% Er. Starting from the ground state $^4I_{15/2}$, the next states are $^4I_{13/2}$ producing the IR peak around 1,540 nm; $^4I_{11/2}$ and $^4I_{9/2}$; $^4F_{9/2}$ being possibly responsible for the red peak around 650 nm; $^4S_{3/2}$ and $^2H_{11/2}$ causing the doublet structure in the green; $^4F_{7/2}$, $_4F_{5/2}$, $^4F_{3/2}$, $^2H_{9/2}$, and $^4G_{11/2}$ contributing in the blue-violet spectral region (a basic energy level scheme of Er^{3+} is given in Fig. 5.23, Sect. 5.4.2). The line positions of possible transitions from these excited states to the ground state are given in Fig. 4.11. The FWHM of the $^4I_{13/2} \rightarrow {}^4I_{15/2}$ transition is only 15 meV, which is less than the crystal field splitting of the $^4I_{15/2}$ state that ranges between 25 and 65 meV depending on the crystal type of the host [233]. In contrast to this, the FWHM of the peaks in the visible range is much larger achieving values in the order of 200 meV. While the power

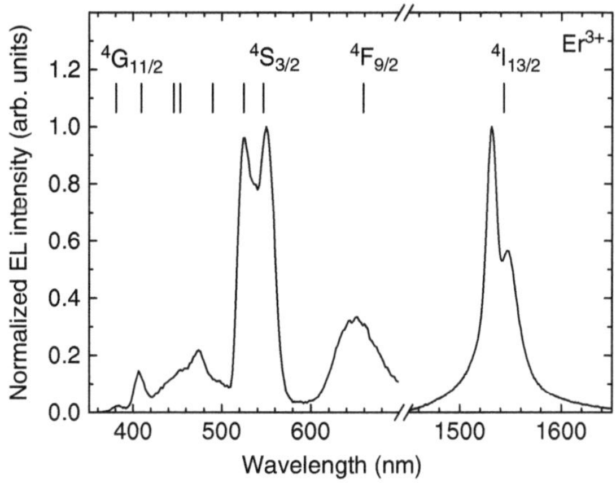

Fig. 4.11 Typical EL spectrum of Er-implanted MOSLEDs. The spectra in the visible and the IR were recorded by different detectors

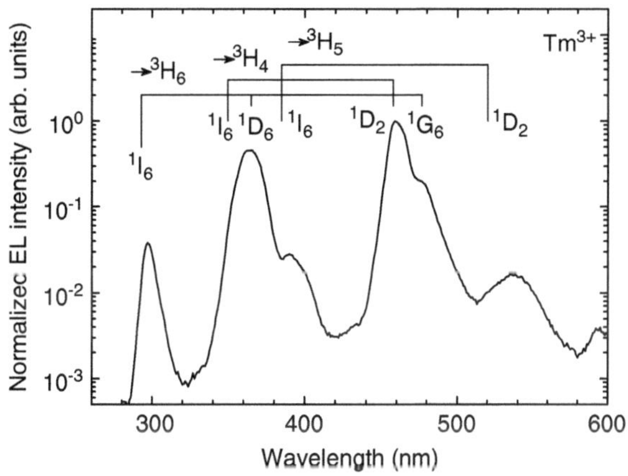

Fig. 4.12 Typical EL spectrum of Tm-implanted MOSLEDs

efficiency in the IR can reach the 10^{-3} range, it hardly exceeds a value of 10^{-5} in the visible. Therefore, defect luminescence from ODCs or NBOHCs can significantly contribute to the observed EL in the blue and red spectral region, respectively. EL decay times in the range of 2.1 and 4.4 ms were measured for the $^4I_{13/2} \rightarrow {}^4I_{15/2}$ transition [260].

4.1.4.7 Thulium

In the periodic system of elements, the neighbour of Er to the right is Tm with now 12 electrons in the $4f$ shell. The EL spectrum as shown in Fig. 4.12 is dominated by a blue peak at 460 nm and two UV satellites around 364 and 297 nm. In addition, there is also a minor sub-peak at around 540 nm. The main peaks at 460 and 364 nm seem to be composed of several components, and indeed the observed spectrum can be explained by the possible transitions from the higher lying states $^1I_6$, $^1D_2$ and $^1G_6$ to the ground state sub-levels $^3H_6$, $^3H_4$ and $^3H_5$. Both the energetic

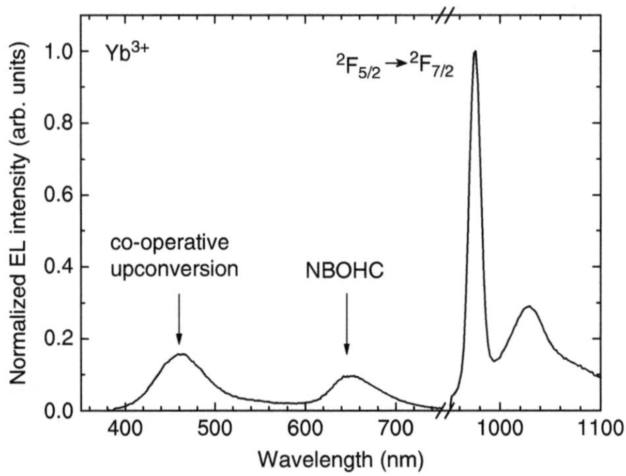

Fig. 4.13 Typical EL spectrum of Yb-implanted MOSLEDs. The spectra in the visible and the IR were recorded by different detectors

distance between the ground state sub-levels as well as their crystal field splitting is large, so, for example the crystal field splitting of the $^3H_6$ state amounts to 100 meV for Tm^{3+} in Y_2O_3 [233]. EL power efficiencies in the 10^{-5} range were measured.

4.1.4.8 Ytterbium

Yb^{3+} lacks one electron in order to have a filled $4f$ shell, for which reason there is only one ground state with $L = 3$ and $S = 1/2$ that splits similarly to Ce^{3+} into the levels $^2F_{5/2}$ and $^2F_{7/2}$, but now with $^7F_{7/2}$ being the lowest level. Interestingly, the EL spectrum of Yb-implanted SiO_2 layers as shown in Fig. 4.13 exhibits more EL lines as expected. The IR peaks at 975 nm (FWHM 14 nm) and 1,025 nm are assigned to the radiative transition $^7F_{5/2} \rightarrow {}^7F_{7/2}$ [147], and the Stark splitting of the $^7F_{7/2}$ level, which can amount up to 75 meV depending on the host material [233], may account for the observed double peak structure. The blue peak at 475 nm is most probably due to a co-operative upconversion process in which two neighbouring Yb^{3+} ions being in the $^7F_{5/2}$ state simultaneously relax to the $^7F_{7/2}$ state and emit only one photon [261, 262]. The EL decay time of the 475 nm line is with 1.3 ms approximately half the value of the EL decay time of the IR peaks with 2.3 ms [147], which is a strong indication for the co-operative upconversion process [261]. The EL peak at 650 nm is assumed to be due to NBOHCs. In the present case, the EL power efficiency is weak compared to most of the other RE-implanted SiO_2 layers.

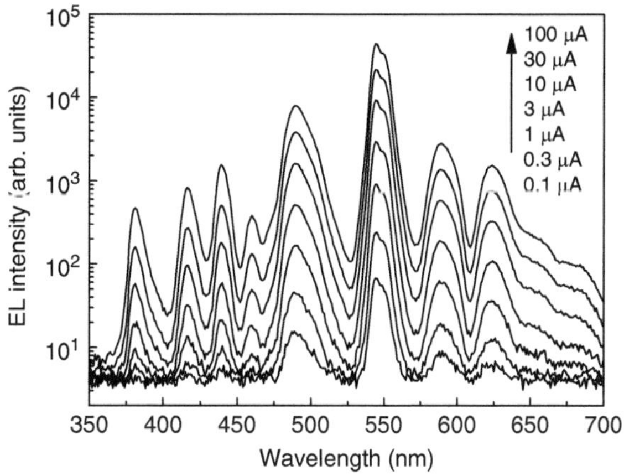

Fig. 4.14 EL spectrum of Tb-implanted MOSLEDs at different injection currents. The device area is $7 \times 10^{-4}\,\mathrm{cm}^2$

4.2 Injection Current Dependence

4.2.1 The Excitation Cross Section

Figure 4.14 displays the EL spectrum of Tb-implanted MOSLEDs at different injection currents. On a first look, it seems that the shape of the spectrum is independent of the specific electric excitation conditions and that the EL intensity shows a linear dependency on the injection current density. Both properties are advantageous in respect of potential applications and – as it will be shown in the following – are good approximations for a broad range of fabrication and operation conditions. Larger deviations from this linearity occur for very low currents, very high currents, or if the EL spectrum is composed of EL lines originating from different excited $4f$ levels of the RE^{3+} ion as in the case of Eu.

As shown in Fig. 4.15a, the linear dependency between EL intensity and the injection current density j is limited to a range of approximately three orders of magnitude. The EL power efficiency obtained from this data has a plateau of nearly constant efficiency roughly between 10^{-5} and $10^{-2}\,\mathrm{A\,cm^{-2}}$ (Fig. 4.15b). There is a slight descent with increasing j as the applied voltage slowly increases with increasing j (IV curves of Sect. 3.2.1). At lower injection current densities, lower electric fields have to be applied, which lead to an energy distribution of the hot electrons in the CB of SiO_2 (Sect. 3.1.2) where the fraction of electrons having enough energy to excite the RE^{3+} ions is small. With increasing j, higher electric fields have to be applied, and in turn, the electron energy distribution shifts to higher energies, enhancing the fraction of electrons with sufficient high energies. As a result, a superlinear rise of the EL-j dependence for low injection current densities is observed. Approaching medium and high injection current densities, the fraction of high-energetic electrons still increases, but relatively slowly. Simultaneously, first saturation effects take effect on a small scale. The balance between these two processes leads to the wide range of linearity, which ends at very high injection current densities when saturation becomes dominant.

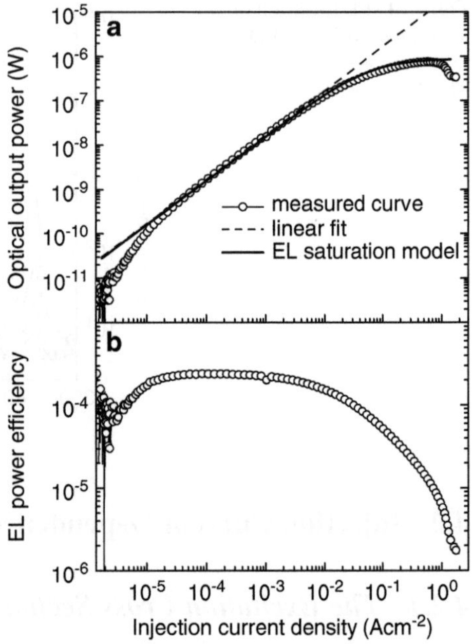

Fig. 4.15 EL output power of a Tb-implanted MOSLED as a function of the injection current density (**a**). The same data were plotted as power efficiency vs. injection current density (**b**)

The saturation of the EL can be understood with the assumption that there is a limited number of RE^{3+} ions that can be excited only once within a time interval in the order of the EL decay constant. This EL saturation model was applied not only to Ge-implanted MOSLEDs [101] but also to other EL devices such as Er in Si [117]. A quantitative description can be derived by considering the rate equation of a two-level system. If N_{max} and N^* denote the total number of excitable and the number of really excited RE^{3+} ions, the latter one changes with time according to

$$\frac{\mathrm{d}N^*(t)}{\mathrm{d}t} = \frac{\sigma_{\text{EL}} j}{q}\left[N_{\text{max}} - N^*(t)\right] - \frac{N^*(t)}{\tau} \tag{4.1}$$

with σ_{EL}, j, q and τ being the EL excitation cross section, the injection current density, the elementary charge and the EL decay time, respectively. The first and second term on the right side of (4.1) describe the current-induced excitation rate and the overall decay rate, respectively. The solution gives

$$N^*(t) = A\exp\left[-\left(\frac{\sigma_{\text{EL}} j}{q} + \frac{1}{\tau}\right)t\right] + C \quad \text{with} \quad C = N_{\text{max}}\frac{j}{j + q/\sigma_{\text{EL}}\tau}. \tag{4.2}$$

Under steady state conditions ($t \to \infty$), the observed EL intensity should obey the relation

$$\text{EL} \propto \eta_{\text{IQE}}\frac{N^*}{\tau} \tag{4.3}$$

with η_{IQE} being the internal quantum efficiency (Sect. 5.1.1). By combining (4.2) and (4.3), the EL–j dependency can be modelled by

$$\frac{EL}{EL_{max}} = \frac{j}{j + q/\sigma_{EL}\tau}. \tag{4.4}$$

As shown in Fig. 4.15a, the EL saturation can be described quite well with this model. If τ is known from EL decay time measurements, the product $\sigma_{EL} \cdot \tau$ can be fitted from the EL–j dependence in order to obtain the EL excitation cross section σ_{EL}. Typical values for σ_{EL} are in the range of 10^{-15} to 10^{-14} cm^2.

4.2.2 *Colour Shift of the Electroluminescence Spectrum*

In the last chapter, it was shown that the onset of a certain EL line occurs when the fraction of electrons with energies higher than the required excitation energy of this line exceeds a certain threshold value. If the EL spectrum is composed of EL lines originating from different excited states of the RE ion, the threshold for the lower lying excited states is achieved at lower electric fields than the threshold of the higher lying ones. Lower fields mean lower injection current densities, for which reason the EL intensity ratio between EL lines from different excited states depends on the injection current density.

The practical consequences can be studied in case of Tb. For the sake of simplification, the EL lines originating from the $^5D_3$ and $^5D_4$ excited state are called the "blue" and "green" lines, respectively, although the corresponding spectral regions extend up to the UV (for $^5D_3$) or to the red (in case of $^5D_4$). In Fig. 4.16a, the EL intensity for the $^5D_3 \rightarrow {}^7F_5$ (413 nm) and $^5D_4 \rightarrow {}^7F_5$ (541 nm) line is plotted vs. j for a Tb concentration of 2%. Obviously, the EL–j dependency from the higher lying state $^5D_3$ starts at much higher injection current densities, which is the equivalent of higher electric fields. This late onset is not only due to the lower EL intensity of the 413 nm line as demonstrated in Fig. 4.16b. In this graph, the EL intensity ratio between the blue and the green EL line, called the B/G ratio, is displayed as a function of the applied electric field for different Tb concentrations [126]. All dependencies show an increase with increasing electric field in accordance with the aforementioned model, which assumes a shift of the energy distribution of the electrons in the CB of SiO_2 to higher energies. This in turn alters the excitation probability in favour of higher energetic states. The change of the B/G ratio changes the colour impression, which becomes clearly visible by the naked eye if the B/G ratio crosses or is at least close to one. Furthermore, the EL intensity ratio of two lines that originate from the same excited state, for example that of the $^5D_4 \rightarrow {}^7F_5$ (541 nm) and $^5D_4 \rightarrow {}^7F_4$ (589 nm) line, does not change with j or the applied electric field. The concentration dependence also visible in Fig. 4.16b is the subject of Sect. 4.3.

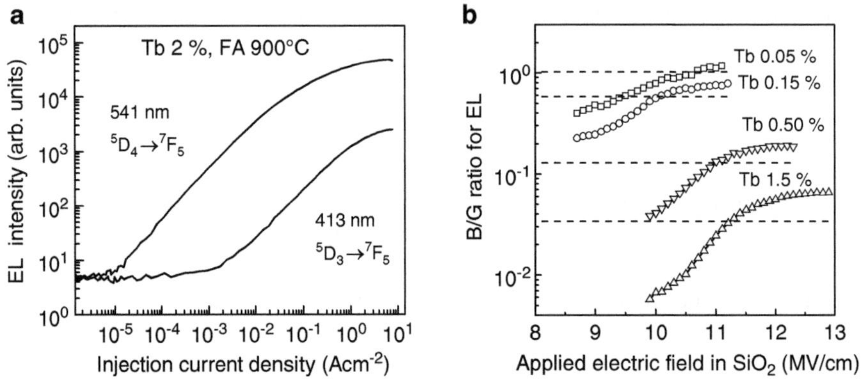

Fig. 4.16 EL – j dependence for the $^5D_3 \rightarrow {^7F_5}$ (413 nm) and $^5D_4 \rightarrow {^7F_5}$ (541 nm) line for a Tb-implanted MOSLED device (**a**) and dependence of the B/G ratio (see the text for definition) on the average electric field for samples with different Tb concentrations (**b**). The *horizontal lines* show the values of the B/G ratio of the PL spectra using an excitation wavelength of 240 nm (after [126])

This colour shift is much more pronounced in case of Eu, but here the physical interpretation is complicated by the interference of additional processes such as quenching. As mentioned in Sect. 4.1.4.3, the EL spectrum of Eu-implanted SiO_2 is mainly composed of a narrow EL band around 618 nm due to a $^5D_0 \rightarrow {^7F_2}$ transition and a broad EL band in the blue, which is believed to be caused by a $5d \rightarrow 4f$ transition within the Eu^{3+} and Eu^{2+} ion, respectively [128]. The EL–j dependency of these two EL bands, the blue one represented by a prominent peak at 400 nm, is shown in Fig. 4.17. For low injection current densities, only the red EL increasing linearly with the injection current density is observed. For intermediate currents, the blue EL is also observed with smaller intensity, but obviously with a slope larger than one. Indeed, if a power law dependence $\mathrm{EL} = j_0 + A \cdot j^n$ is assumed, an exponent $n = 1.4$ can be fitted up to $1\,\mathrm{A\,cm^{-2}}$. Deviations from the model discussed in Sect. 4.2.1, especially in the linear region, can be generally interpreted in such a way that the excitation cross section changes with j. The physical background for such a behaviour can be the change of the excitation conditions (e.g. a change in the local electric field) or a change of the number of excitable luminescence centres (e.g. by blocking luminescence centres caused by charge trapping). A value of $n = 1.4$ indicates an increase of the excitation cross section with increasing charge injection. At a current density of $0.02\,\mathrm{A\,cm^{-2}}$, the slope of the red EL–j curve drops down to $n = 0.45$, which cannot be modelled by (4.4). Without giving further details, such a phenomenon can be explained by the onset of a quenching mechanism. Because of the larger slope, the 400 nm EL exceeds the red EL for current densities larger than $0.15\,\mathrm{A\,cm^{-2}}$ before another quenching mechanism diminishes strongly the 400 nm EL above $1\,\mathrm{A\,cm^{-2}}$.

Such a strong injection current dependence gives a more drastic change in colour than in case of Tb. Taking the ratio between the blue and red EL intensity, the curve in Fig. 4.18 is obtained [256]. Depending on the injection current a red, blue or

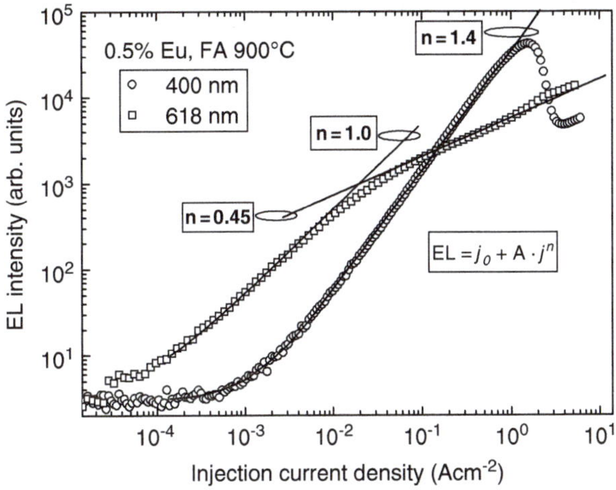

Fig. 4.17 EL intensity vs. injection current density for 0.5% Eu and FA 900°C. The *solid lines* are fits assuming a power law dependence

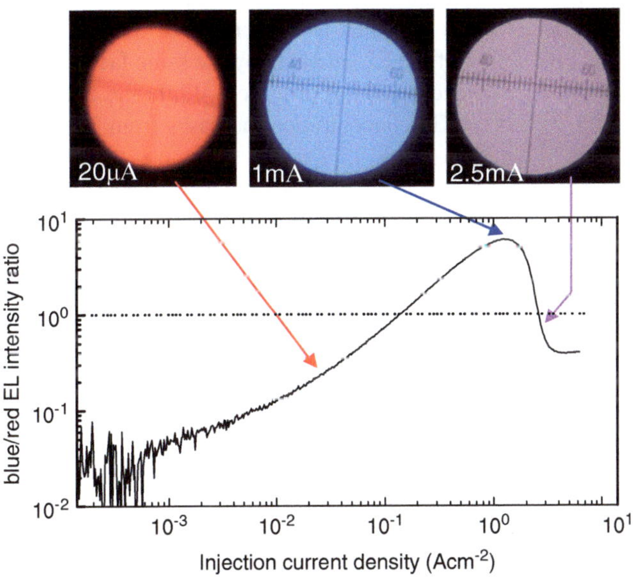

Fig. 4.18 Ratio between the blue and red EL lines of the Eu spectrum as a function of injection current. The photographs show the same device at different injection currents (after [256])

violet EL well visible by the naked eye can be observed, whereby the violet is naturally a mixed colour. In principle, this would allow the construction of a multi-colour device where the colour depends on the injection current [128, 263]. Note that the exposure times of the photographs are different depending on the brightness of the EL.

4.3 Concentration Quenching

4.3.1 The General Case

In general, the EL intensity increases with increasing RE concentration as long as a maximum is reached at a certain concentration, followed by a decrease of the EL intensity for even higher concentrations. The EL increase at low concentrations can be simply understood as the increasing number of luminescence centres with increasing RE concentration. At higher concentrations, quenching and saturation effects dominate the EL concentration dependency. This behaviour is well illustrated in Fig. 4.19a, showing the internal and external quantum efficiency of Gd-implanted SiO_2 layers as a function of Gd concentration. For a Gd concentration between 1.5 and 3%, a maximum of EL is achieved, before the efficiency drops down considerably. While the EL decay time (Fig. 4.19b) exhibits a similar development as the EL efficiency, the excitation cross section saturates for high Gd concentrations (Fig. 4.16c).

The interaction between neighbouring luminescence centres, two neighbouring Gd^{3+} ions in this case, is a common quenching mechanism that leads to an increase of non-radiative decay paths and thus to a decrease of the EL decay time. This should result in a reduction of the EL intensity by the same factor as the decay time was shortened which, however, was not observed. While the decay time becomes shorter by about 30% if the Gd concentration increases from 3 to 9% (Fig. 4.19b),

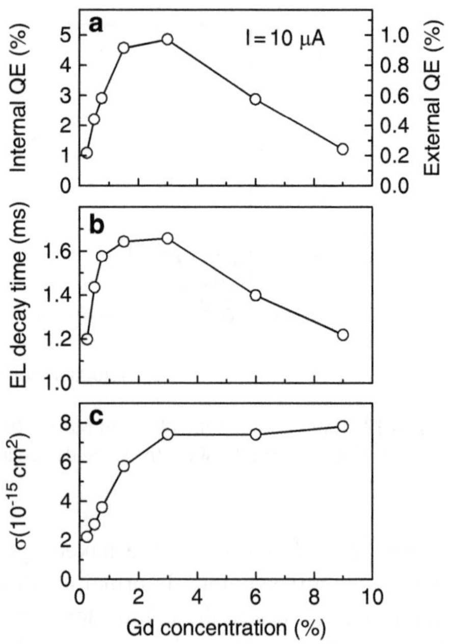

Fig. 4.19 Internal and external quantum efficiency (**a**), EL decay time (**b**) and excitation cross section (**c**) for different Gd concentrations of Gd-based MOSLEDs (after [135])

the quantum efficiency drops down by a factor of 5 (Fig. 4.19a) not including the fact that the amount of potentially available Gd increases by a factor of 3 at the same time. Therefore, the increase of non-radiative decay paths caused by the interaction between neighbouring Gd^{3+} ions gives only a minor contribution to the observed EL quenching.

It should be noted that there is another mechanism that can lead to a smooth *decrease* of the EL decay time and a concurrent *increase* of the EL intensity with increasing cluster size. According to Fermi's golden rule, the spontaneous radiative emission rate is proportional to the final density of states, which is enhanced in the vicinity of dielectric interfaces [264]. LCs being located close to or at the surface of REO_x clusters may take advantage of this effect, which is the higher the larger the size and the refraction index[1] of the clusters. If the non-radiative decay paths remain unchanged, an enhanced radiative transition rate would increase the EL intensity and decrease the EL decay time. However, the intensity of this effect is assumed to be in the order of 10–20%.

Because of the constant excitation cross section between 3 and 9% Gd (Fig. 4.19c), the only possibility to explain the decrease of the EL intensity is the assumption that the number of excitable LCs decreases. There are at least three different mechanisms that can contribute to this decrease. At first, it is reasonable to assume that the number and – to a lesser extent – the size of the GdO_x clusters increase with increasing Gd concentration, although the concentration dependence of the cluster growth was not explicitly investigated. As a result, the fraction of the Gd ions located within a GdO_x cluster is increasing, too. Unfortunately, such a location is very unfavourable with respect to electron impact excitation, as the impinging electron may excite Gd^{3+} at the surface of the GdO_x cluster, but is strongly decelerated when entering the cluster. This point will be discussed in more detail in Sect. 4.3.3. Second, the amount of oxygen available for the oxidation of Gd is limited, for which reason the fraction of Gd that can achieve a trivalent ionic configuration Gd^{3+} also decreases. Finally, the number of defects and electron traps is higher for higher Gd concentrations (Fig. 3.20). If a Gd^{3+} ion is close enough to a negatively charged trap, it can be screened by Coulomb repulsion, which prevents the excitation of this Gd^{3+} ion by hot electrons. Thus, the fraction of Gd^{3+} ions facing Coulomb screening may also rise with increasing Gd concentration.

The optimum concentration depends on the RE element and the annealing conditions. So the EL spectrum of Ce-implanted MOSLEDs annealed with FA 900°C shows interestingly only small variations (<10%) in the concentration range of 0.5 and 3%. A close inspection of the EL–*j* dependence (Fig. 4.20) reveals in fact a close match of the curves but different EL saturation levels: with increasing Ce concentration, the saturation is achieved at lower and lower EL levels. This tendency can be quantified by fitting the curves with (4.3), and the result is given in the inset of Fig. 4.20. The maximum EL intensity EL_{max}, which is proportional to the number of

---

[1] Please not that the dielectric constant $\varepsilon \approx n^2$ (for $\mu \approx 1$) is similar in nature but not in value to the ε used in equations 3–6 ff. as these values apply to completely different frequency ranges: to optical frequencies ($\sim 10^{15}$ Hz) and to the low frequency limit (electrical DC conditions).

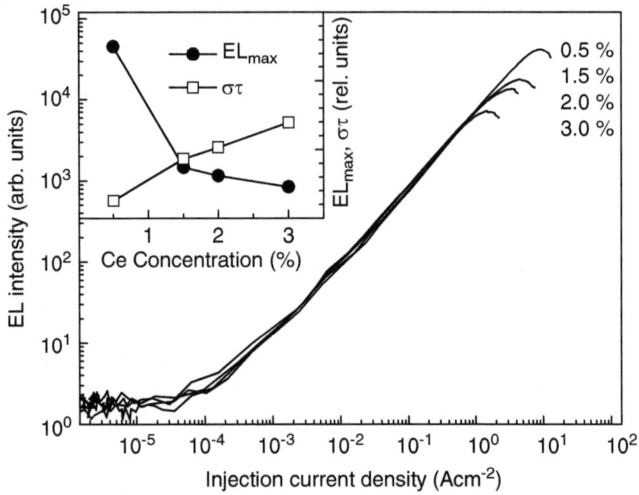

Fig. 4.20 EL – j dependency for different Ce concentrations and FA 900°C. The *inset* shows the relative change of the maximum EL intensity and the $\sigma_{EL} \cdot \tau$ product with Ce concentration

excitable Ce^{3+} ions, clearly drops down by a factor of 5. However, in order to meet the experimental observation and keep the EL spectrum unaffected of this, the $\sigma_{EL} \cdot \tau$ product has to rise at the same time. In the present case, only a decay time of 75 ns for 1% Ce is known, so that it cannot be decided whether the rise of $\sigma_{EL} \cdot \tau$ is due to a rise of σ_{EL} or τ. Nevertheless, $\sigma_{EL} \cdot \tau$ is in the 10^{-20} range leading to a cross section in the order of $10^{-13}\,cm^{-2}$. It has to be noted that the upper part of the EL–j curves is interfered by quenching effects, leading to a sagging of the EL dependency instead of a convergence to the saturation value at very high injection currents. Moreover, a slight non-linearity can be found for currents below 100 μA. If a power law dependency EL $\propto j^n$ is fitted to the EL dependency, an exponent $n \approx 0.9$ is obtained, which indicates the presence of a weak EL quenching mechanism already at low injection currents.

Another example is Er, where the EL intensity of the green double peak (500 and 550 nm) and the prominent IR peak at 1.54 μm is plotted vs. the Er concentration (Fig. 4.21) for FA 900°C and RTA 1,050°C. The green EL peak has a maximum around 0.8%, whereas the maximum intensity of the IR peak is achieved for concentrations higher than 1.4% [120] only. Note that the visible and the IR parts of the EL spectrum were recorded by two different detectors, for which reason the intensity scales in Fig. 4.21a, b are not comparable.

4.3.2 Cross Relaxation

One type of EL quenching belonging to the category of mechanisms that are based on neighbour–neighbour interaction is cross relaxation. The impact of this

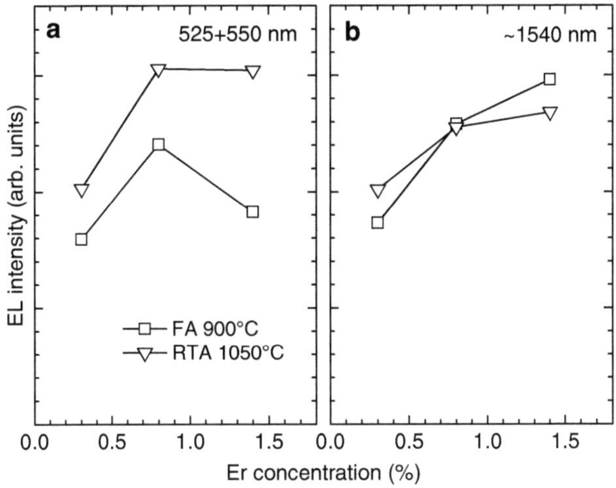

Fig. 4.21 EL intensity of the *green double peak* (**a**) and the IR peak at 1.54 μm (**b**) for different Er concentrations and two annealing procedures. The visible and the IR parts of the EL spectrum were recorded by two different detectors

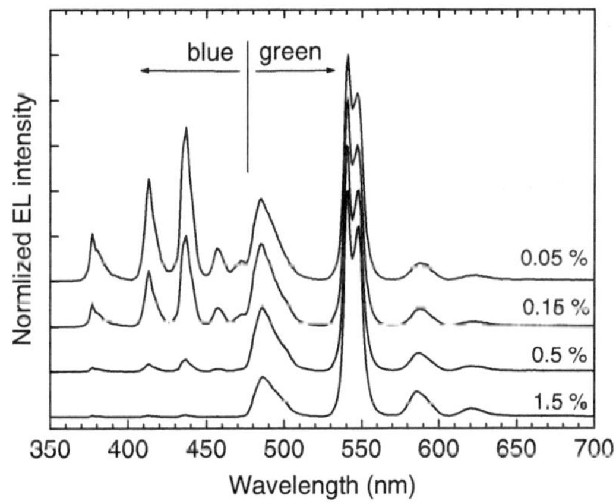

Fig. 4.22 EL Spectra of Tb-implanted SiO_2 layers with different Tb concentrations. The structures were annealed with FA 800°C (after [126])

mechanism can be studied in Fig. 4.22 comparing the EL spectra of Tb-implanted MOSLEDs for different Tb concentrations. The spectra are normalized in respect of the EL peak at 541 nm, and obviously the "blue" EL lines that originate from the excited $^5D_3$ state are quenched in comparison to those resulting from the $^5D_4$ state [126].

The quenching of the blue EL is quantified in Fig. 4.23a showing the EL intensity of the green $^5D_4 \rightarrow {}^7F_5$ and blue $^5D_3 \rightarrow {}^7F_5$ EL line as a function of the Tb concentration. The intensity of the green line increases linearly with the Tb concentration up to 2% Tb and starts to saturate for higher concentrations. In contrast, the intensity of the blue line reaches a small maximum at a concentration of 0.15% Tb followed by a smooth decrease for higher concentrations. So the intensity ratio

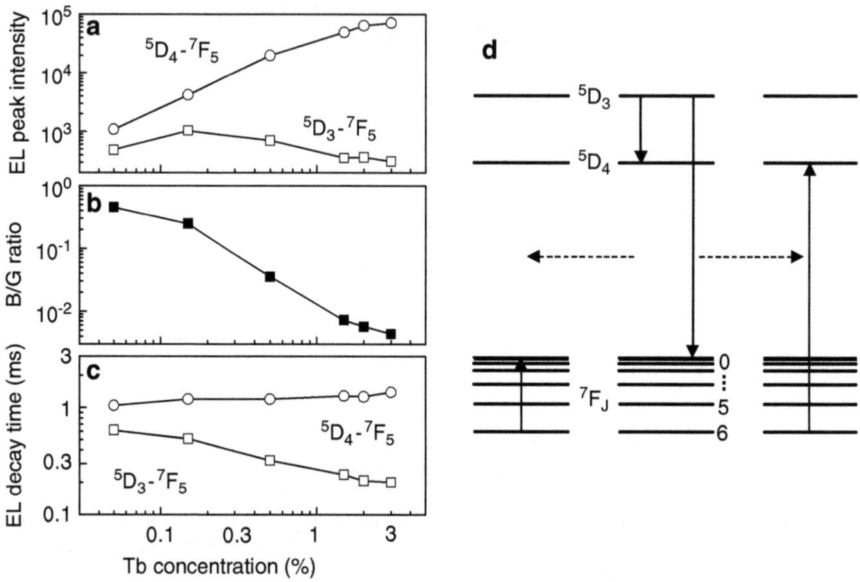

Fig. 4.23 Tb concentration and annealing dependence of the EL intensity of the *blue* and *green* EL (**a**), their intensity ratio (**b**) and their EL decay time (**c**). On the *right side*, the cross relaxation modes of two neighbouring Tb^{3+} ions are schematically shown (**d**) (after [126, 256])

between the blue and the green EL line (B/G ratio) drops down from an initial value of 0.45 for 0.05% Tb by two orders of magnitude to 0.004 for 3% Tb (Fig. 4.23b).

The investigation of the EL decay time (Fig. 4.23c) revealed that the decay time for the blue line decreases continuously with increasing Tb concentration, whereas the decay time for the green line slightly increases. This indicates that with increasing Tb concentration, a non-radiative decay path affecting only the $^5D_3$ level becomes more and more intense. This behaviour is ascribed to the cross-relaxation from the $^5D_3$ to the $^5D_4$ level in neighbouring Tb^{3+} ions as it was observed in Tb-doped yttrium silicate glass [265]. As sketched in Fig. 4.23d, cross relaxation becomes possible due to the small energy mismatch of 11 meV for $E(^5D_3 - {}^5D_4) \approx E(^7F_0 - {}^7F_6)$. In the first case, a Tb ion being in the $^5D_3$ level ends up non-radiatively in the $^5D_4$ level, and the transferred energy is used to excite a neighbouring Tb ion from the ground state $^7F_6$ to the $^7F_0$ level, respectively. Analogous to this, the non-radiative transition $^5D_3 \rightarrow {}^7F_0$ excites a Tb ion being in the ground state into the $^5D_4$ level. Because of the small energy distances, $^7F_0$ can be also replaced by $^7F_1$ or even $^7F_2$. As a result, the B/G ratio decreases.

However, it has to be noted that cross-relaxation is not the only quenching mechanism for the blue EL. While the blue intensity drops down by two orders of magnitude, the decay time reduces only by a factor of 3. Other principal possibilities to explain the quenching of an EL line is the reduction of the number of luminescence centres or the reduction of the excitation cross section. As both the green and blue EL originate from the Tb^{3+} ion, the first possibility can be excluded. A reduction of the excitation cross section can be understood by considering that with increasing Tb concentration the defect concentration in the SiO_2 layer increases, too. This in turn increases the probability of inelastic scattering of hot electrons leading to a reduction of their average kinetic energy and thus their ability to excite high-energetic states of the Tb^{3+} ion. Comparing two structures with a low and a high concentration of Tb, in the latter case, the fraction of hot electrons that have enough energy to excite Tb^{3+} into the $^5D_3$ level is lower than for low Tb concentrations, whereas the excitation into the lower lying $^5D_4$ level is less affected by this.

4.3.3 Europium: The Interplay Between Di- and Trivalent Ions

As already mentioned, the EL spectrum of Eu-implanted SiO_2 is very sensitive to the fabrication conditions. Such spectra are displayed in Fig. 4.24 for different Eu concentrations showing a great variability in the blue-green spectral region. According to Sect. 4.1.4.3, the blue-green EL is mainly due to an electronic transition of an electron from the *5d* to the $4f$ shell of an Eu^{2+} ion with minor contribution of ODCs and luminescence centres in the SiON layer. The *5d* orbital is sensitive to the chemical environment, which qualitatively explains the variability in the blue-green as the SiO_2 microstructure changes remarkably with Eu concentration.

Before the variability of the EL spectrum is correlated with microstructure, the main tendencies should be extracted. For this purpose, the integral EL intensity over the EL lines caused by Eu^{3+} is divided by the integral over the entire spectrum. This ratio, called the red EL fraction in the following, is plotted in Fig. 4.25 as a function of the injection current density for different Eu concentrations (a) and as a function of the Eu concentration for different annealing procedures (b). The injection

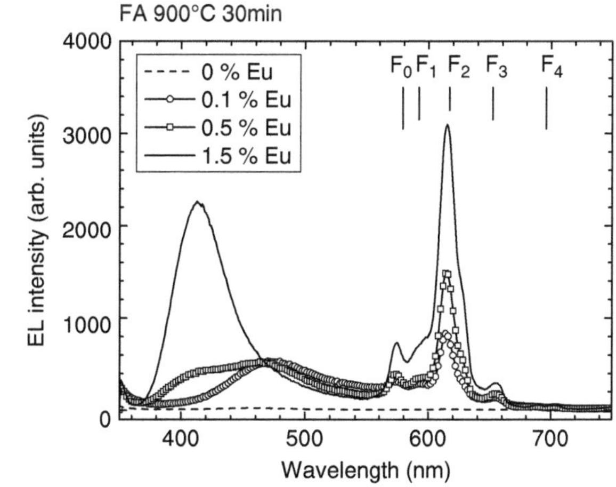

Fig. 4.24 EL spectrum of Eu-implanted MOSLEDs annealed with FA at 900°C for various Eu concentrations and under an injection current of 5 μA (7.1×10^{-3} A cm^{-2}). The *solid lines* indicate the position of the $^5D_0$–^7F_J lines ($0 \leq J \leq 4$) according to [233] (figure after [131])

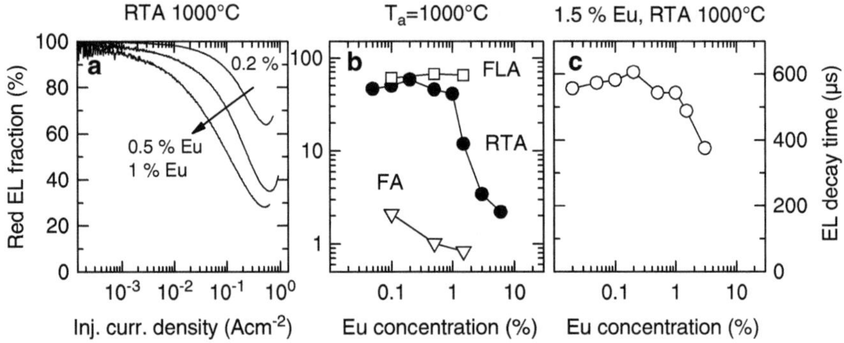

Fig. 4.25 The dependence of the relative intensity of the Eu^{3+} emission in % (red EL fraction) on the injection current (**a**) and the Eu concentration (**b**). Note the different scales in (**a**) and (**b**). The EL decay time as a function of Eu concentration for RTA 1,000°C is given in (**c**). The data were obtained using an injection current density of 7×10^{-3} A cm^{-2}

current dependence itself is already known from Sect. 4.2.2, but with increasing Eu concentration, the red EL fraction drops down more heavily. The same tendency can be observed for FA 1,000°C and RTA 1,000°C in Fig. 4.25b; only for FLA, the red EL fraction changes only slightly with Eu concentration. The decrease of the red EL fraction does not need to be real quenching; as demonstrated in Fig. 4.24, the absolute intensity in the red still increases with Eu concentration, but not so strong like in the blue. Indeed, the question if the red is really quenched depends on annealing: for RTA 1,000°C, the reduction of the red EL fraction goes along with a decrease of the total EL intensity. Figure 4.25b also reveals that the concentration and annealing dependence in case of Eu are closely interweaved, but in order to keep the presentation as structured as possible, the annealing dependence is shifted to the next chapter.

Searching for a better insight, the first indication is the EL decay time dependence on Eu concentration, which is displayed for RTA 1,000°C in Fig. 4.25c. The EL decay time of the red $^5D_0$–$^7F_2$ line smoothly decreases from 606 μs for 0.2% Eu to 375 μs for 3% Eu, for which reason the interaction between neighbouring Eu^{3+} ions cannot account for the drastic drop down of the red EL fraction in Fig. 4.25b for concentrations above 1%. Once again the differences in microstructure for different RE concentrations play the key role for interpretation, but in contrast to Tb, there are now two different luminescence centres to be considered, namely Eu^{2+} and Eu^{3+}. The observed preference for Eu^{2+} for high Eu concentrations is therefore closely linked to the question, under which circumstances and in which local environment the Eu atom can achieve a trivalent ionic configuration. As shown in Sects. 2.2.3 and 2.3, the microstructure of Eu-implanted devices strongly depends on the preparation conditions, and there are at least three different processes involved [190].

First, during implantation, Si and oxygen are released from the SiO_2 network either by nuclear collisions or by electronic energy deposition leading to bond breaking (Sect. 2.2.1). In the following phase of annealing, the released oxygen can be used to oxidize Eu (Sect. 2.3.2). The second process is the preference of EuO compared to Eu_2O_3. Although the formation enthalpy of Eu_2O_3 with $-1{,}650\,kJ\,mol^{-1}$ (that is $-825\,kJ\,mol^{-1}$ per Eu atom) is higher than that of EuO with $-565\,kJ\,mol^{-1}$ [188], the supply of oxygen that is available for the oxidation of Eu determines whether EuO or Eu_2O_3 dominates. At low Eu concentrations, a single Eu atom can find enough oxygen in its local environment to achieve a Eu_2O_3 configuration. With increasing Eu concentration, the average distance between neighbouring Eu ions decreases, and a competition of the Eu ions for oxygen starts, which finally shifts the weight from Eu^{3+} to Eu^{2+} for high Eu concentrations.

The third development to be considered is the formation and the subsequent ripening of Eu oxide clusters. Similar to the results for Gd (Sect. 4.3.1), the number and size of EuO_x clusters will increase with increasing Eu concentrations. Because of the higher dielectric constant of Eu oxide with $\varepsilon(Eu_2O_3) \approx 22$ [266] compared with $SiO_2(\varepsilon \approx 3.9)$ and the continuity of the dielectric displacement, the electric field within an Eu oxide cluster is lower than in the SiO_2 matrix. Lower electric fields are not high enough to accelerate electrons to energies needed for the excitation of Eu ions, for which reason Eu ions inside a cluster are assumed to be optically inactive. Hot electrons moving in the conduction band of SiO_2 are able to excite Eu ions, which are located at the surface of an Eu oxide cluster only. This argumentation can be expanded to other RE oxides as they have comparably high ε values [266]. The values tabulated for different RE oxides in [267] range from 12 to 15. In general, large REO_x clusters decrease the EL intensity and can shift the weight to the blue EL in case of Eu assuming that large clusters contain more Eu^{2+} than small clusters. However, due to the low number of experimental indications (Sect. 2.3.2), the latter point remains speculative.

Considering these three processes, the shift of the colour impression from red to more blue for higher Eu concentrations is mainly due to a decrease of the concentration ratio between available oxygen and Eu. In addition, for a given annealing procedure, higher Eu concentrations tend to form larger clusters, which are believed to favour the blue emission.

4.4 Annealing Dependence

4.4.1 Short Time vs. Long Time Annealing

As already mentioned, annealing is an important processing step in order to anneal out defects induced by ion implantation or other processing steps and to activate LCs. Moreover, annealing can improve the quality of the SiON layer, which was deposited at a temperature of about 325°C. However, high temperatures trigger diffusion driven processes such as NC formation and the out-diffusion of the implanted species. Both processes decrease the number of active luminescence centres and thus the EL efficiency, as RE ions being located at the SiO_2–Si interface or in

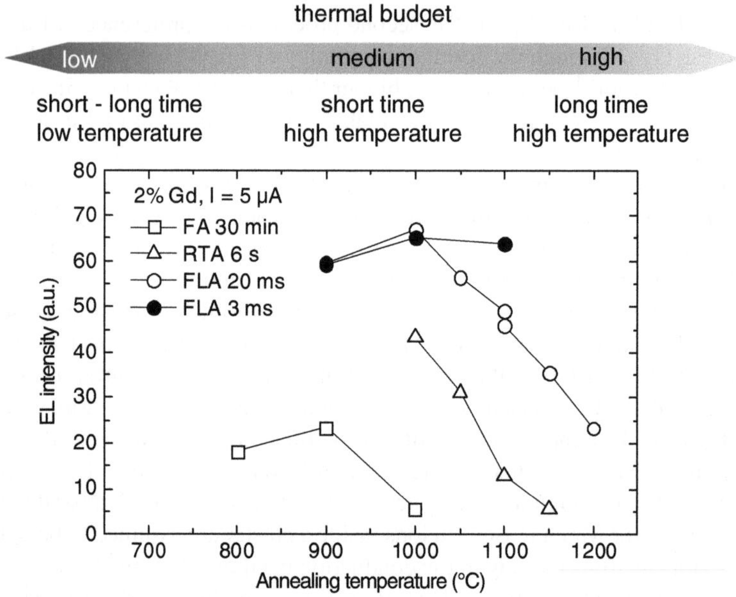

Fig. 4.26 EL intensity as a function of annealing temperature for Gd-implanted MOSLEDs and for different annealing types. The EL intensity was measured at 316 nm under an injection current of 7×10^{-3} A cm^{-2} (original data from [138, 139])

the core of a NC are assumed to be optically inactive. There should be an optimum between a low thermal budget (low temperatures), where defects are not yet annealed out completely, and a high thermal budget (high temperatures and long annealing times).

Figure 4.26 gives an overview of the annealing dependency of MOSLEDs implanted with 2% Gd. Although the temperature range of the different types of annealing covers only a narrow part of the temperature scale, the general tendency can be clearly seen: at low temperatures, the EL intensity increases, reaches a maximum between 900 and 1,000°C and decreases for even higher temperatures. The optimum temperature shifts slightly to higher temperatures with shorter annealing times but depends also on the implanted RE element and its concentration. More importantly, the EL intensity increases with decreasing annealing time. Therefore, optimum annealing conditions will normally be achieved for medium thermal budgets, which means high temperatures and short annealing times. The physical reason for these optimum conditions is mainly the fact that in most cases the annealing of defects requires a local rearrangement of the electronic structure such as bond formation or relaxation. Such processes are often characterized by an activation energy, which has to be provided by applying a sufficient high temperature. On the other hand, these processes are short-range processes taking place on a short time scale. In contrast to this, diffusion driven processes are long-range processes taking place on a fairly long time scale. For the sake of comparison, it should be noted that for

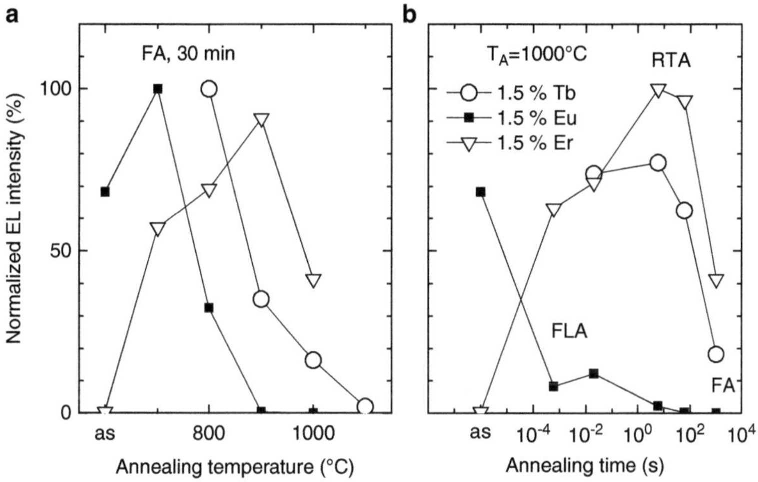

Fig. 4.27 Normalized EL intensity of the main Tb^{3+} ($^5D_4 \rightarrow {}^7F_5$), Eu^{3+} ($^5D_0$–$^7F_2$) and Er^{3+} ($^4I_{13/2}$–$^4I_{15/2}$) line of RE-implanted MOSLEDs on the annealing temperature (**a**) and the annealing time (**b**) for various types of annealing. The intensities are normalized to the maximum EL of the respective element

FA, the decay time increases smoothly from 1.35 ms to 1.79 ms if temperature rises from 800°C to 1,000°C [135]. This tendency is opposite to the expected behaviour of EL quenching by non-radiative decay paths, for which reason this weak decay time dependence is less important for the physical interpretation of the annealing time dependence of the observed EL.

The EL dependency of Gd can be found in a similar form for other RE elements, but with different positions of the EL maximum on the annealing temperature or annealing time scale. Figure 4.27 displays the normalized EL intensity of the main line of Tb-, Eu- and Er-implanted MOSLEDs as a function of the different annealing types and annealing temperatures. In case of Eu, maximum EL occurs only for low thermal budgets (FA 700°C), whereas this maximum is shifted to higher thermal budgets in case of Tb and Er (FA 800–900°C, RTA 1,000°C 6 s). The Eu behaviour is most probably caused by the higher diffusivity of Eu, leading to the formation of large clusters which – according to Sects. 4.3.3 and 4.4.2 – is connected with a much lower and more bluish EL.

Finally, a short comment is devoted to stability. In general, the stability in terms of time to breakdown under constant current injection increases with increasing thermal budgets. So the devices annealed with FLA 1,200°C, RTA 1,100°C or FA 1,000°C feature the highest stability which, however, is of little use as the EL intensity has fallen to unacceptable low values. For low thermal budgets, the EL intensity is often not too bad with values about 50% of the maximum EL intensity for the as-implanted state (e.g. Eu in Fig. 4.27a) which, however, is accompanied by a stability decrease by more than one order of magnitude (Sect. 6.1.2). Therefore, the best compromise between stability and efficiency is often at medium thermal budgets, for example FA 800–900°C or FLA/RTA 1,000°C.

4.4.2 Spectral Shifts with Cluster Evolution

In the same way, as changes in the RE concentration cause *spectral shifts* due to a selective (or favoured) excitation of luminescence centres, different annealing conditions have a similar effect due to the microstructural development, namely the cluster growth. Figure 4.28 displays the annealing temperature dependence of MOSLEDs with 1.5% Tb under an injection current density of 0.014 A cm^{-2}. The rise of the annealing temperature from 800°C to 1,100°C results in a strong quenching of both the green $^5D_4 \rightarrow {}^7F_5$ and blue $^5D_3 \rightarrow {}^7F_5$ EL line (Fig. 4.28a), while the B/G ratio as defined in Sect. 4.3.2 increases (Fig. 4.28b). As the decay time of the green EL line is more or less constant (Fig. 4.28c), the quenching cannot be due to a non-radiative decay path, which increases with annealing temperature. Quite contrary to this, the increase of the decay time for the blue EL line indicates a *reduction* of the non-radiative decay path responsible for the quenching of the blue EL line.

With increasing annealing temperature, the REO_x cluster size strongly increases (Sect. 2.3.1). As argued in Sect. 4.3.3, Tb atoms agglomerating in clusters contribute no longer to the EL resulting in an effective decrease of the number of LCs and thus a decrease in EL intensity. The increase of the B/G ratio can be understood assuming that the major part of the EL signal originates from Tb^{3+} ions dispersed in the SiO_2 matrix. With increasing annealing temperature, their number decreases due to cluster growth and out-diffusion, especially for FA 1,100°C (Fig. 2.10). This in turn increases the average distance between the dispersed Tb^{3+} ions, resulting in a decrease of the quenching caused by cross-relaxation. As a consequence, the B/G ratio and the decay time of the blue EL line increase [256].

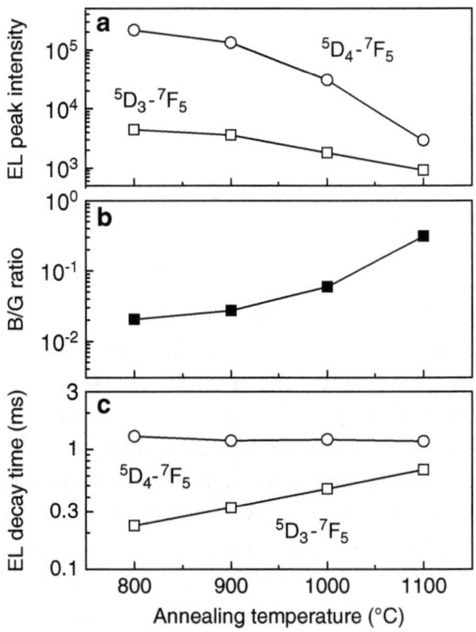

Fig. 4.28 Annealing dependence of the intensity of the Tb-based *blue* and *green* EL line (**a**), their intensity ratio (**b**) and their EL decay time (**c**) (after [126, 256])

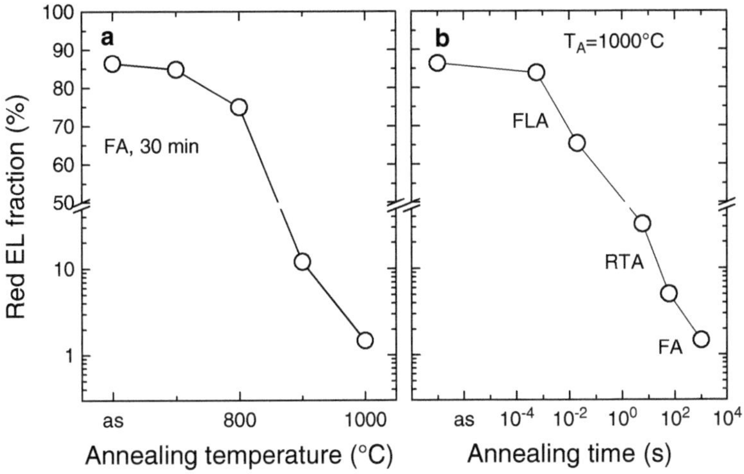

Fig. 4.29 The dependence of the relative intensity of the Eu^{3+} emission in % of Eu-implanted MOSLEDs on the annealing temperature (**a**) and the annealing time (**b**) for various types of annealing. Note the break of the ordinate

In Fig. 4.29, the red EL fraction of Eu-implanted MOSLEDs – defined as the ratio between the red EL intensity caused by Eu^{3+} and the total EL intensity (Sect. 4.3.3) – is plotted as a function of the annealing temperature (a) and the annealing time (b) for various types of annealing. Clearly, this red EL fraction drops down both with increasing annealing temperature and time, and thus the colour impression shifts from red to blue. According to Fig. 2.14, the cluster size strongly increases with increasing thermal budget. Similar to Sect. 4.3.3, larger clusters are rather associated to Eu^{2+} than to Eu^{3+}, which would explain the annealing dependence shown in Fig. 4.29. As the total EL intensity also decreases with increasing thermal budget (but not as strong as the Eu^{3+} intensity), the weak blue EL for high thermal budgets might overlap with the EL from other defects like ODCs (Sect. 4.1.3).

Chapter 5
Electroluminescence Efficiency

The electroluminescence efficiency of a light emitter is of special interest as this property is a critical parameter for most of the potential applications. Indeed, a lot of research on Si-based light emission is devoted to the EL efficiency enhancement of the corresponding light emitters. Therefore, an independent chapter is devoted to efficiency and related issues, in which the focus is not so much on a detailed description of efficiency properties but more on the different strategies to enhance the efficiency of a specific RE-implanted device. Whereas Sect. 5.1 discusses some general aspects of efficiency, Sect. 5.2 reflects general dependencies of the efficiency, namely those on oxide thickness, host material and the implanted RE element. The strategies for efficiency enhancement outlined in the following are based on the co-implantation of other elements, and consequently, Sect. 5.3–5.5 deal with RE-implanted light emitters co-doped with group IV elements, other RE elements or fluorine.

5.1 General Considerations

5.1.1 Definition of Efficiency

The EL efficiency is not only of interest because of its impact on potential applications but is also closely linked with the underlying mechanisms of luminescence creation, and, moreover, is a popular measure to compare the results of different scientific groups among each other. There are three different types of efficiency frequently found in literature, namely the internal quantum efficiency (IQE) η_{IQE}, the external quantum efficiency (EQE) η_{EQE} and the power efficiency η_{PE}.

The IQE is the efficiency closest to the underlying physical principles, and is defined as the ratio between the number of emitted photons per time unit of a single luminescence centre and the number of particles exciting it in the same time interval. Basically, it is the probability that a luminescence centre, once excited, is emitting a photon. Thereby the exciting particles are photons or electrons in case of PL or EL, respectively. However, the IQE neglects all processes preventing an excitation of the LC and all processes in which the emitted photon is not coupled out like re-absorption. The EQE considers these processes but ignores the difference between the energy of the exciting particle and the energy of the emitted photon. It regards the light emitter device as a black box and defines its value as the ratio between the photons emitted by this black box and the number of particles entering it. Finally, the power efficiency calculated by the optical output power divided by the optical or electrical input power covers now all relevant processes and is that type of efficiency which really matters for application. Moreover, it is the easiest type of efficiency to measure. In fact, in many cases where an IQE or EQE is given, their values are calculated from an experimentally measured power efficiency. The different types of efficiency obey the following relation:

$$\eta_{\mathrm{IQE}} > \eta_{\mathrm{EQE}} > \eta_{\mathrm{PE}}, \tag{5.1}$$

and the power efficiency is the only one which cannot exceed a value of 1 in the present case of EL. This can be easily realized by imaging that one electron passing the MOS structure can excite more than one LC if the oxide layer is thick enough. If not otherwise stated the term "efficiency" is used synonymously to power efficiency throughout the book.

In some cases, the fraction of excited luminescence centres f is a better quantity to compare different light emitters than the different types of efficiency. f is defined as the ratio between the number of excited and the total number of luminescence centres, and for RE-implanted MOSLEDs it is given by

$$f = \frac{P_{\mathrm{opt}}\tau}{E_{\mathrm{Ph}}DA}. \tag{5.2}$$

Hereby, the term $P_{\mathrm{opt}}\tau / E_{\mathrm{Ph}}$ with the optical power P_{opt} emitted by the device, the EL decay time τ and the photon energy E_{Ph} gives the number of excited LCs, whereas the product between the implantation dose D and the device area A is equal to the total number of LCs. The underlying assumptions are that each implanted RE ions is at least a potential LC and that an excited LC is excited statistically one time in the time interval τ. Equation (5.2) can be easily extended to light emitters which were fabricated without ion implantation by replacing D by the product of atomic concentration and layer thickness. P_{opt} can be expressed by the product $\eta_{\mathrm{PE}} \cdot P_{\mathrm{el}}$ with η_{PE} and P_{el} as the measured power efficiency and the electric input power, respectively. Finally, (5.2) can be read as

$$f = \frac{\tau V_0 \eta_{\mathrm{PE}} j}{E_{\mathrm{Ph}} D}, \tag{5.3}$$

with V_0 and j as the applied gate voltage and the current density j at this voltage, respectively.

5.1.2 *Efficiency Measurement*

The experimental access is simple as long as a higher precision is not needed. By placing a calibrated Si photodiode or any other kind of optical power meter in a known distance r in a sufficiently dark ambient, the total optical power P_{opt} emitted by the device can be calculated by

$$P_{opt} = \frac{1}{2}\frac{2\pi r^2}{A_{Det}} P_{Det}, \tag{5.4}$$

with P_{Det}, A_{Det} and $2\pi r^2$ being the optical power directly measured by the detector, the detector area and the area of a half sphere with radius r around the light emitter, respectively. Furthermore, the light emitter area is supposed to be small compared to the detector area. As shown in Fig. 5.1, the light emitters, independent if implanted with Ge or RE elements, behave like a Lambertian radiator where the angular EL intensity distribution follows a $\cos\beta$ law with β as the angle between the surface normal of the emitter area and the observation direction. The integral of $\cos\beta$ over the upper half space yields the factor 1/2. Clearly, this method is an integral measurement over the entire EL spectrum within the detector range, but with the known EL spectrum and the detector function the efficiency of single EL peaks can be estimated. The uncertainties of the efficiency values obtained by this method and given in this book are guessed to be a factor of 2 at maximum. This value can be decreased to about 20% by reducing the uncertainties introduced by the geometry (uncertainty of r or slight misalignments), but for significantly more precise measurements, advanced techniques using an Ulbricht sphere etc. have to be used.

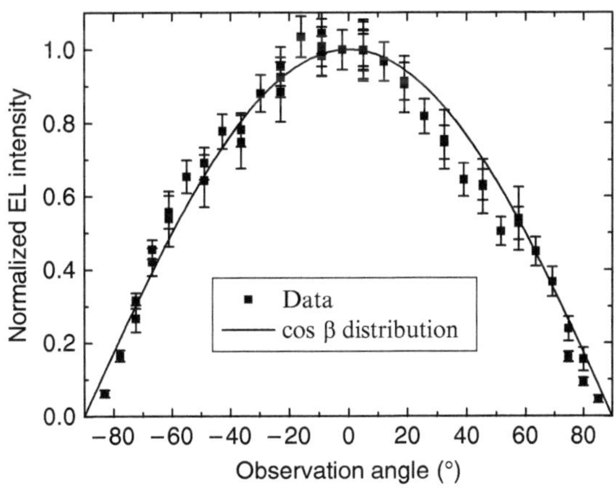

Fig. 5.1 Directional characteristic of a Ge-implanted MOSLED. The smooth indentation at the *right* is the shadowing of the contact needle

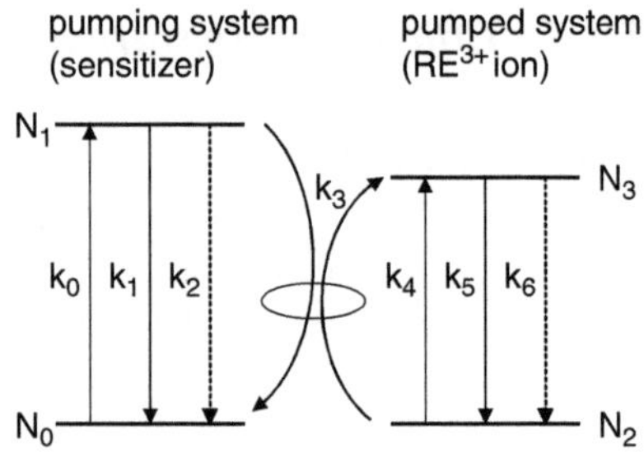

Fig. 5.2 Basic energy scheme of a 2-level system pumping another 2-level system. Radiative transitions are marked by *solid* and non-radiative ones by *dashed arrows* (after [297])

5.1.3 Pumping of a Two-Level System

As pumping of RE ions is a major strategy to enhance the light emitter efficiency, it will support the following discussion to have a closer look to the general situation where a pumping 2-level system, called the sensitizer, is transferring energy to a second 2-level system. Without loss of generality, the latter one is denoted as RE^{3+} ion in the following. The basic energy level scheme of the two systems is depicted in Fig. 5.2 with N_i as the number of sensitizers or RE^{3+} ions being in the ground state or the excited state of interest according to Fig. 5.2. k_i denote radiative (solid arrows) and non-radiative (dashed arrows) transitions between the different levels. The transitions k_0 and k_4 are radiative in case of PL and non-radiative in case of EL. k_3 symbolizing the energy transfer has an exceptional position as it links the two systems via a joint transition rate and has therefore a different physical unit [see (5.5)].

The aim of the following mathematical considerations is to get the time-dependent population of the excited states N_1 and N_3 in order to discuss the changes in EL intensity and EL decay time in case of pumping (Sect. 5.3 and 5.4). Similar considerations can also be found throughout the literature, for which reason, the representation is limited to the minimum extent which is necessary to follow it in a comfortable manner.

The change of N_1 and N_3 with time can be described by the rate equations

$$\left.\begin{aligned}\dot{N}_1 &= k_0 N_0 - k_1 N_1 - k_2 N_1 - k_3 N_1 N_2 \\ \dot{N}_3 &= k_3 N_1 N_2 + k_4 N_2 - k_5 N_3 - k_6 N_3\end{aligned}\right| . \tag{5.5}$$

In order to get a simple analytical solution, the approximation of weak excitation is introduced implying that much more RE^{3+} ions are in the ground state N_2 than in the excited state N_3. This allows the relation $N_2 \approx N_{RE}$ with N_{RE} being the total number of RE^{3+} ions *coupled* to the sensitizer. This approximation is equal to the condition $k_4 \ll k_5 + k_6$ and disentangles the two rate equations (5.5) which can be read now as

$$\left.\begin{aligned}\dot{N}_1 &= k_0 (N_S - N_1) - k_1 N_1 - k_2 N_1 - k_3 N_1 N_{RE} \\ \dot{N}_3 &= k_3 N_1 N_{RE} + k_4 N_{RE} - k_5 N_3 - k_6 N_3\end{aligned}\right| , \tag{5.6}$$

with N_S being the total number of coupled sensitizers. The first equation can be solved straightforward with the usual ansatz of an exponential function

$$N_1(t) = A + B\exp(-\alpha t). \tag{5.7}$$

The corresponding conditions for rise and decay of an excitation pulse give the following constants for the EL decay:

$$A = 0; \quad B = \frac{k_0 N_S}{\alpha + k_0}; \quad \alpha = k_1 + k_2 + k_3 N_{RE}. \tag{5.8}$$

Note that α differs by the summand k_0 depending if the rise or the decay phase of a pulse is considered. From (5.8) it immediately follows that the decay time of the sensitizer shortens in case of pumping according to the increase of α. If k_0 is small, the EL intensity of the sensitizer decreases by the same factor; for higher k_0 values, the decrease is less pronounced. However, in real cases, it may happen that there are two groups of sensitizers: those coupled and those not coupled to the RE^{3+} ions. The EL contribution of the un-coupled sensitizers, independent if EL intensity or EL decay time is considered, remains unchanged, and only the values for coupled sensitizers will decrease. If the contribution of both groups, coupled and un-coupled sensitizers, is in the same order, the measured EL decay time characteristic becomes bi-exponential. However, if k_3 is large, the EL signal from the coupled sensitizers can be so weak that its contribution to the EL decay time vanishes. In this case, no shortening of the EL decay time but a decrease of the EL intensity is observed by a factor equal to the ratio of the coupled to the total number of sensitizers.

Equation (5.7) can be inserted in the second equation of (5.6), and the bi-exponential ansatz

$$N_3(t) = C + D\exp(-\alpha t) + E\exp(-\beta t) \tag{5.9}$$

gives the solutions for the rise and the decay under pulsed excitation. The steady-state intensity under pumping is

$$k_5 N_3^R(\infty) = \frac{k_5 N_{RE}\,(B\,k_3 + k_4)}{\beta}, \tag{5.10}$$

with N_3^R being the number of excited RE^{3+} ions under rise conditions. Obviously, the intensity under pumping is higher than that without pumping because of the additional term Bk_3 in the numerator. The boundary conditions for the decay deliver

$$\begin{aligned} C = 0; \quad D &= \frac{k_3 N_{RE} B}{(\beta - \alpha)}; \quad E = \frac{N_{RE}\,(Bk_3 + k_4)}{\beta} - D \\ \alpha = k_1 + k_2 + k_3 N_{RE}; \quad \beta &= k_5 + k_6. \end{aligned} \tag{5.11}$$

The first case to be considered is that the energy transfer is much faster than the intrinsic decay of the RE^{3+} ion which seems to be not too uncommon as the decay time of $4f$ intrashell transitions of RE^{3+} ions is usually in the ms range. The corresponding approximation is $\alpha \gg \beta$, and a comparison of the constants E and D reveals that $E \gg D$, for which reason the EL decay of the pumped RE^{3+} ions is monoexponential with the original time constant $1/\beta$. The situation changes if a relatively slow transfer process is pumping a RE element with a fast transition, as it e.g. occurs for Ce (Sect. 5.4.1). In this case, the decay is bi-exponential, and the question which decay time dominates depends on the special values of the different transition rates.

5.1.4 Strategies for Efficiency Tuning

Generally, there are three different strategies to enhance the EL efficiency:

1. The excitation cross section is enhanced. This implies that with the same electrical input power, a higher number of LCs are excited.
2. If a LC is excited, the probability to emit a photon is maximized. This is equal to maximizing the IQE.
3. The probability of a photon to leave the device after being emitted by a LC is enhanced. This is equal to maximizing the outcoupling efficiency.

All strategies for efficiency enhancement presented in the next chapters belong to the first category. The most common method is the use of sensitizers with a sufficient large cross section which pumps the LCs additionally by transferring energy from the sensitizers to the LCs. In a first approximation, one can say that the excitation cross section of the LCs is expanded by that of the sensitizer. In the following, two types of sensitizers are discussed in more detail, namely, NCs from group IV elements and other RE elements, which are a matter of Sect. 5.3 and 5.4. Fluorine co-doping which is discussed in Sect. 5.5 increases the excitation cross section by decreasing the number of competing defects.

The second type of efficiency enhancement strategies is directly linked with the EL decay time. As found in Chap. 4, the EL decay time varies significantly, but not more than a factor of 2 or 3 with the fabrication conditions, for which reason the room for improvements is limited. Therefore, the main strategy is to select fabrication conditions with a fairly long decay time and to avoid any process or operation mode which reduces the decay time remarkably.

The third category includes techniques which try to suppress the internal back reflection of emitted photons and to enhance the emission in a certain, mostly narrow wavelength region. One example is surface patterning or the shaping of a surface layer in the form of a 2D photonic crystal [269–272]. Following this idea, the active layer of the light emitter itself can be part of a photonic crystal [273]. Other possibilities are the use of Bragg mirrors [274] or plasmonic structures on the surface

[275–277]. However, some of these techniques manipulate the final density of states (Sect. 4.3.1), which, strictly speaking, interferes with the second type of efficiency.

Before the strategies for efficiency enhancement are treated in detail, the dependency of the EL power efficiency on the RE element, the SiO_2 layer thickness and the host matrix are discussed in the following chapter.

5.2 Geometry and Material Aspects

5.2.1 Comparison Between Different Rare Earth Elements

Figure 5.3 summarizes the power efficiencies η_{PE} measured for different RE-implanted MOSLEDs with a SiO_2 thickness of 100 nm, a SiON thickness of 100 nm and a RE peak concentration of 1.5% for different annealing conditions. As seen, the results vary from one RE element to the other. Whereas Tb and Er allow a highly efficient EL in the 10^{-3} range, Ce, Eu and Gd belong to a group of medium efficiency. That of Ce can be improved by Gd co-doping (Sect. 5.4.1), and in case of Gd the measured η_{PE} values are probably underestimated because of the absorption of the 316 nm line in ITO which is not exactly known. For Eu, good efficiencies can only be achieved in a narrow bandwidth of the fabrication conditions. Tm, Yb and Pr (not shown) exhibit efficiencies which exceed the level of unimplanted SiO_2 layers only by a factor <10.

Generally, the results are a good guide to evaluate the potential for possible applications, but they are qualitative in nature. Because of the large spectrum of

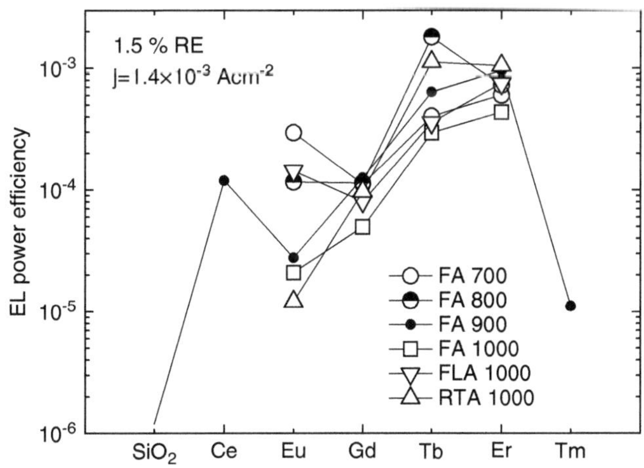

Fig. 5.3 EL power efficiency of different RE-implanted MOSLEDs and for different annealing procedures. Unimplanted MOSLEDs are marked by "SiO_2". The values for Er mainly cover the IR peak about 1,540 nm

fabrication conditions, only a limited range could have been investigated, and there might exist more favourable fabrication conditions especially for the RE elements of low efficiency.

5.2.2 The Dark Zone Model

As there are different potential applications with different voltage specifications, the question how much the operation voltage of a Si-based light emitter can be scaled down is of special interest. Electrons which are injected into the CB of SiO_2 are accelerated and reach an equilibrium energy distribution in a distance between 10 and 20 nm behind the injecting interface (Sect. 3.1.2 or [206]). Electrons of the high energy tail of this distribution have enough energy to excite RE^{3+} ions by impact excitation. This implies that electrons being closer to the injecting interface have no enough energy, for which reason a dark zone behind the injecting interface can be postulated in which RE^{3+} ions are not able to exhibit luminescence due to insufficient excitation conditions. Whereas this fact might be of minor importance for thick SiO_2 layers, it becomes critical if the SiO_2 thickness is scaled down.

In order to verify the existence of such a zone and to estimate its extension, a set of Tb-based light emitters with decreasing SiO_2 thickness was investigated [278] whose fabrication conditions together with some measured EL properties are given in Table 5.1. In a first step, EL power efficiencies were measured at a current density of 1.4×10^{-4} A cm^{-2} (0.1 μA), which is a value low enough to ensure the absence of any saturation effects and high enough to be unaffected by background noise. The corresponding EL spectra are more or less identical to the EL spectrum shown in Fig. 4.10. In a second step, the fraction of excited luminescence centres f according to (5.3) was estimated from the power efficiency assuming an EL decay time of $\tau \approx 1.3$ ms [126], and both values are given in Fig. 5.4 as a function of the SiO_2 layer thickness. The smaller power efficiency at 100 nm compared to 75 nm is due to the fact that the applied voltage increased with increasing layer thickness, but without an adequate increase of the Tb implantation dose. f gives similar values at

Table 5.1 Device and performance parameter of Tb-implanted light emitters with different SiO_2 layer thicknesses.

Layer thickness SiON/SiO_2 (nm)	Implantation Dose (ions per cm^2)	Implantation Energy (keV)	R_P (nm)	Measured power efficiency η	V_0 (V)	Fraction of excited Tb ions f	Dark zone width (nm)
100/100	3×10^{15}	100	48	6.7×10^{-4}	141	1.58×10^{-5}	
100/75	3×10^{15}	50	37	9.6×10^{-4}	91	1.46×10^{-5}	
100/50	2×10^{15}	35	24	5.7×10^{-4}	63.5	9.1×10^{-6}	24
100/30	1.4×10^{15}	20	17	1.2×10^{-4}	48.5	2.1×10^{-6}	16

R_P and V_0 denote the projected range of implantation and the applied gate voltage, respectively. In this sample set, the Tb implantation is performed first, followed by SiON deposition

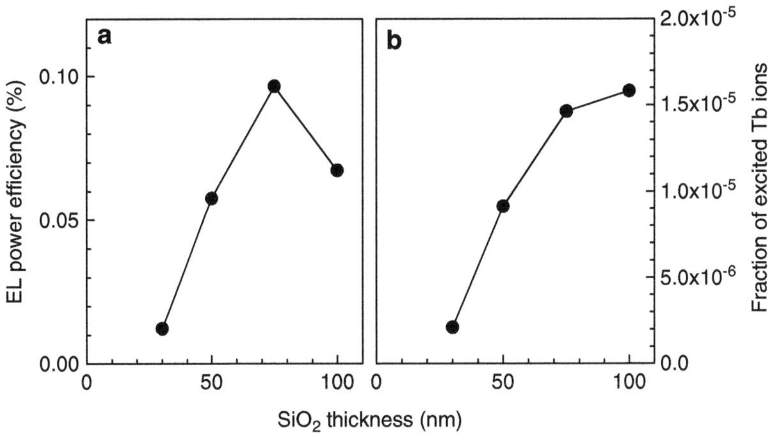

Fig. 5.4 EL power efficiency (**a**) and the fraction of excited Tb^{3+} ions (**b**) as a function of the SiO_2 layer thickness (after [278])

about 1.5×10^{-5} for 75 and 100 nm thick SiO_2 layers followed by a strong decrease for thinner SiO_2 layers. In case of saturation, f can reach the 10^{-2} range.

The influence of the dark zone can now be quantified by the assumption that the drop down in the fraction of excited Tb ions for thin SiO_2 layers is due to the fact that more and more Tb^{3+} ions will be located in the dark zone. If the implantation profile is known, an estimation of the dark zone extension can be done giving the values of 24 and 16 nm for 50 and 30 nm of gate oxide, respectively. The situation is illustrated in Fig. 5.5 showing from top to down the Tb profile according to SRIM simulations [151] for devices with 30, 50, 75 and 100 nm SiO_2. The dark zone is symbolized by a grey rectangle with an experimentally estimated extension for the thinner SiO_2 layers and an assumed extension of 20 nm for 75 and 100 nm thick SiO_2 layers. In the framework of the given experimental uncertainties, this value fits well with theoretical predictions done by Fischetti et al. [206]. Of course, the dark zone interface to SiO_2 is not sharp due to the fact that the energy distribution of the hot electrons is broad.

Although the problem can be somewhat alleviated by using a shallow Tb implantation, it seems to be difficult to obtain efficient MOS-based light emitters for gate oxide thicknesses below 20 nm. On the quest for an alternative approach circumventing this problem, an obvious solution could be the replacement of SiO_2 by another host matrix which is the matter of the next chapter.

5.2.3 *The Influence of the Host Matrix*

In order to investigate the suitability of SiON as an alternative host material for RE ions, a set of Tb-based light emitters was prepared where RE implantation was

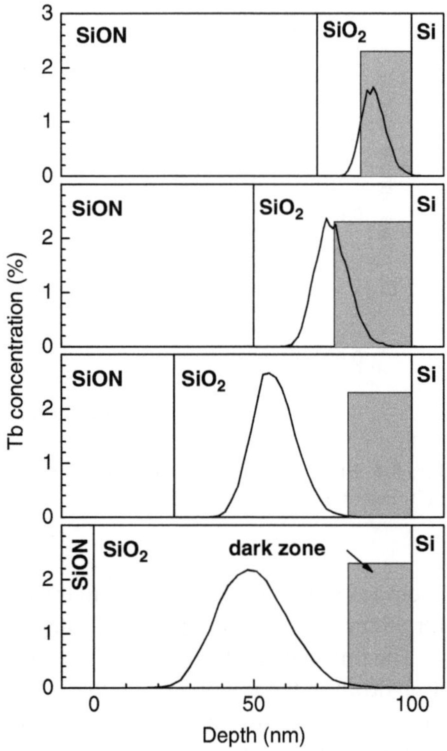

Fig. 5.5 Profile of the implanted Tb ions according to SRIM simulations [151] for devices with 30, 50, 75 and 100 nm SiO_2. The *dark zone* is symbolized by a *grey rectangle*. Its width was estimated from the power efficiency data for 30 and 50 nm (after [278])

Table 5.2 Device and performance parameter of Tb-implanted light emitters with different locations of the projected range R_P

Device notation	Layer thickness $SiON/SiO_2$ (nm)	Implantation		R_P (nm)
		Dose (ions per cm^2)	Energy (keV)	
D1	100/100	3×10^{15}	500	150
D2	100/100	2.2×10^{15}	350	102
D3	100/100	1×10^{15}	140	54
D4	100/10	1×10^{15}	140	54

In this sample set, the SiON deposition is performed *first*, followed by Tb implantation

performed *through* the SiON layer. Energy and dose of this implantation was varied in such a way that the projected range of implantation was located in the middle of the SiO_2 layer, in the interface region between SiO_2 and SiON or in the middle of the SiON layer. The fabrication conditions of the different devices are summarized in Table 5.2.

Figure 5.6a shows the EL power efficiency of Tb-implanted MOSLEDs with different locations of the Tb profile as a function of the injected charge. The corresponding Tb profiles as calculated by SRIM [151] are given in Fig. 5.6b. Note that device D4 has the same Tb profile in SiON as D3 but only a 10 nm thick SiO_2

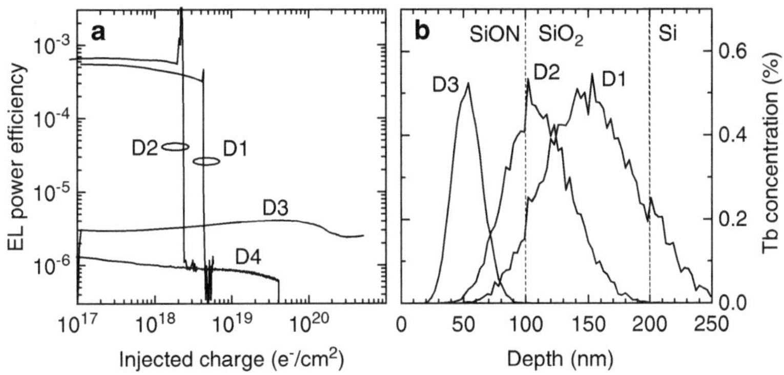

Fig. 5.6 EL power efficiency of Tb-implanted MOSLEDs with different locations of the Tb profile as a function of the injected charge (**a**). The corresponding Tb profiles as calculated by SRIM [151] are given in (**b**). Note that device D4 has the same Tb profile in SiON as D3 but only a 10 nm thick SiO_2 layer [278]

layer. It is clearly seen that the devices with Tb located in the SiO_2 layer or in the SiO_2–SiON interface region (D1 and D2) exhibit a high power efficiency which dramatically drops down if the Tb is located entirely in the SiON layer (D3 and D4). Generally, devices where Tb is implanted into or through the SiO_2–SiON interface (D1 and D2) exhibit a lower electric stability visible by the earlier breakdown in Fig. 5.6a. In addition, device D4 with only 10 nm SiO_2 has an even lower EL power efficiency than D3.

In contrast to SiO_2, SiON is characterized by a much higher defect and trap density, for which reason the electron transport is mainly mediated by a trap-to-trap movement. Such electrons are lost for the Tb^{3+} excitation as they move just above the bottom of the CB of SiO_2 and will not gain much kinetic energy. However, it was found [204] that there is a small fraction of electrons in SiON which behave very similar to the hot electrons in SiO_2. As shown in Fig. 5.7, the energy distribution of these electrons has the same shape than that of SiO_2, but with a number which is two orders of magnitude lower. Indeed, this fits well with the observed differences in power efficiency between devices with Tb in SiO_2 (D1 and D2) and those with Tb being entirely located in SiON (D3 and D4). Electrons entering the CB of SiON from the CB of SiO_2 are decelerated due to the high scattering rate, and the fraction of electrons having enough energy to excite Tb^{3+} is continuously decreasing with increasing distance from the SiO_2–SiON interface. This also explains why the power efficiency of the device with Tb-implanted SiON and a 10 nm thick SiO_2 layer (D4) is lower than that with 100 nm SiO_2 (D3). As 10 nm is shorter than the dark zone extension (see previous chapter), the acceleration distance for electrons is not even enough to achieve the average equilibrium energy in SiO_2.

Consequently, SiO_2 can only be replaced by materials where the charge transport is also mainly carried by hot electrons. In addition, the mean free path between subsequent scattering events of an electron should be larger than that in SiO_2 implying

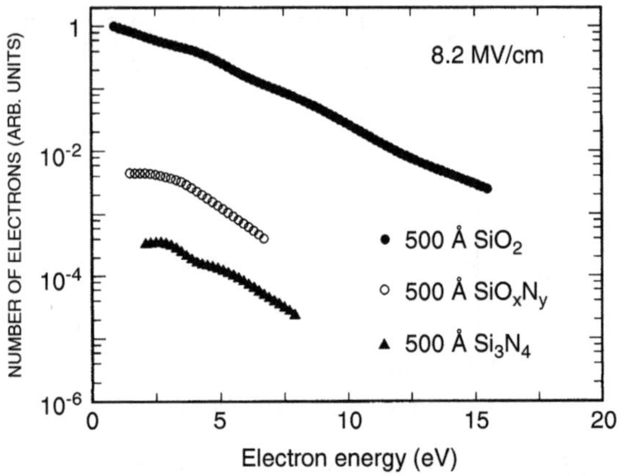

Fig. 5.7 Energy distribution of hot electrons in SiO_2, SiO_xN_y and Si_3N_4 (after [204])

a faster acceleration and a thinner dark zone behind the injection interface. According to Fig. 5.7, Si_3N_4 seems to be even more unfavourable than SiON due to the lower fraction of hot electrons available for impact excitation of RE^{3+} ions.

5.3 Sensitizing by Group IV Nanoclusters

5.3.1 Si Nanoclusters

5.3.1.1 The Energy Transfer Between Si Nanoclusters and Er

One possibility to enhance the EL efficiency is the increase of the EL excitation cross section by group IV NCs, which are assumed to have a much larger excitation cross section than RE^{3+} ions. In fact, it was found that in presence of Si NCs, the PL intensity of the Er^{3+} emission at 1.54 μm can be increased by at least one order of magnitude [118, 260, 279–282]. The Si NCs can be excited by PL with a cross section in the $10^{-17} - 10^{-15}\,cm^{-2}$ range [282–288] which is several orders of magnitude higher than that of Er^{3+} in the $10^{-21} - 10^{-19}\,cm^{-2}$ range [264,284,289,290]. An exciton in the Si NC can recombine non-radiatively transferring the released energy to a nearby Er^{3+} ion which gets excited and thus contributes to the observed PL at 1.54 μm. The energy transfer is evidenced by the increase of the Er-related PL intensity with increasing density of Si NCs (see next chapter, Fig. 5.9), by the decrease of the PL intensity originating from the Si NCs with increasing Er concentration and by the flattening of the PL excitation spectrum of the Er emission with increasing Si content [291]. The latter observation shows that the relatively sharp and resonant direct excitation of Er^{3+} ions transforms into a broad non-resonant excitation via the Si NCs. Doping of Si NCs usually leads to a deterioration of the energy transfer process which, however, can be minimized by a simultaneous P and B doping of the NCs [292].

Detailed investigations of the PL decay time behaviour gave further insights in the energy transfer mechanism. The fact that the decay time of the PL related to Si NCs is independent of the Er concentration is a support for the strong coupling model [280, 283], which implies that the probability of an excited Si NC to transfer its energy to a nearby Er^{3+} ion is much higher than the probability of a radiative de-excitation of the Si NC. This means that Si NCs with a neighbouring Er^{3+} ion will not contribute anymore to the NC-related PL, for which reason the decrease of the Si-related PL intensity with increasing Er concentration is due to a decrease of the number of available LCs (Sect. 5.1.3). However, the decay time of the Er^{3+} luminescence drops down from values around 15 ms for Er in silica [264] to a few ms depending on the Si concentration [280, 291]. This means that on one hand, the Si NCs strongly increase the excitation cross section for Er, but on the other hand, they introduce additional non-radiative decay mechanisms quenching the PL. Time resolved PL measurements in the ns range [293, 294] revealed that there are three different mechanisms determining the PL decay on three different time scales (Fig. 5.8): For times below 1 μs (phase I), a fast decay is observed which is assigned to an Auger quenching process in which the energy of the excited Er^{3+} ion is transferred to an electron in the CB of the Si NC. The subsequent rise of the PL intensity in the range of 1–10 μs (phase II) is due to the energy transfer process to Er^{3+} via exciton recombination followed by the known decay in the ms range (phase III). The same time scale in the order of 1 μs for the energy transfer was already found by Pacifici et al. [284]. Independently, the Auger quenching process was confirmed by PL measurements under current injection [119].

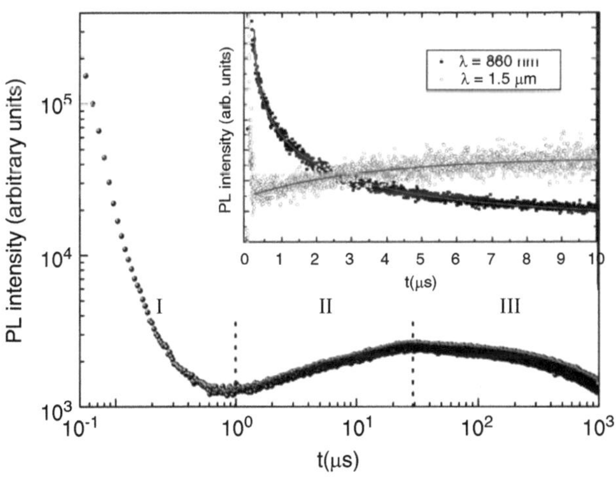

Fig. 5.8 Er-related PL decay at 1.5 μm in Er-doped, Si-rich SiO_2, showing three different temporal phases. The *inset* displays the temporal evolution of the RT PL for the first 10 μs after the excitation pulse, measured at 1.5 μm for Er^{3+} (*open circles*) and 860 nm for Si NCs (*closed squares*). The excitation wavelength is 450 nm (after [293])

Recent investigations [295, 296] show that in case of Si-rich oxide without Si NCs, the energy can be transferred to Er on a ns time scale via ODCs, and that even in case of Si NCs this fast energy transfer process plays a significant role. Recently, the Si=O defect which can be located either in the bulk SiO_2 or at the surface of Si NCs was found to be strongly involved in the energy transfer to Er^{3+} [297].

The discrepancy between the increase of the excitation cross section by several orders of magnitude and the increase of the PL intensity by only 1–1.5 orders of magnitude, both caused by the introduction of Si NCs, implies that only a small fraction of Er^{3+} ions profits from the energy transfer via Si NCs. Indeed, several publications reported a maximum of only ~3% Er^{3+} which can be excited by Si NCs [288, 290]. The maximum interaction distance between a Si NC and an Er^{3+} ion was estimated to be in the order of 0.5 nm only using a distance-dependent interaction model [290] which excludes the majority of Er^{3+} ions from the energy transfer process. Furthermore, the energy transfer process was found to be sequential in nature, i.e., one Si NC can excite only one Er^{3+} ion at a time [283, 284, 286]. To circumvent this problem, newer studies consider the confinement of Si NCs and Er^{3+} in thin layers separated by Al_2O_3 [298] or the use of Si_3N_4 as a host material ([299, 300] and references therein).

5.3.1.2 Electroluminescence

Whereas the introduction of Si NCs into Er-doped SiO_2 causes a strong increase of PL efficiency, the situation looks less optimistic for EL. Figure 5.9 compares the

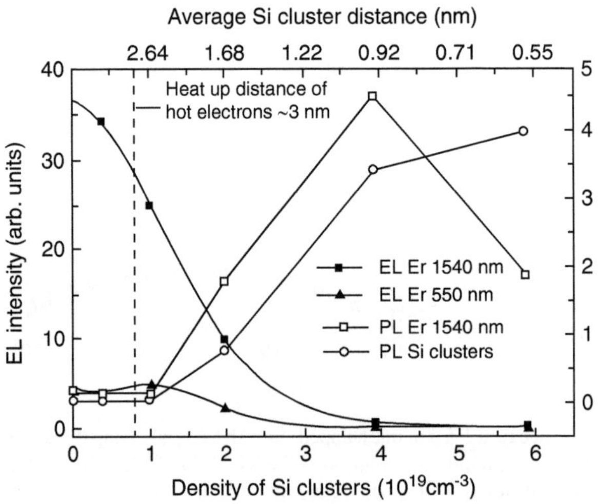

Fig. 5.9 PL and EL intensity of Er-implanted MOSLEDs with different Si concentrations (after [118]). The Si excess concentration increases from 0% to 15% and is in a first approximation proportional to the density of Si clusters. The *dashed line* marks the mean free path of hot electrons in SiO_2

PL and EL intensity of MOSLEDs with a 200 nm thick SiO_2 layer implanted subsequently with Si and Er (after [118]). The Si and Er implantation, the latter one with an Er concentration of 0.24% in the middle of the oxide layer, were followed by FA at 1,100°C, 60 min and by FA 850°C, 30 min, respectively. As the increase of Si concentration results in an increase of Si cluster density rather than an increase of NC size, the Si excess concentration ranging from 0 to 15% can be directly transformed into a Si cluster density. In the present case, an average Si NC size of about 2 nm was roughly estimated from the energetic position of the NC-related EL band using the equation $E_G = E_{G0} + d^{-1.39}$ [301] with E_G, E_{G0} and d being the bandgap energy of the Si NC in eV, the bandgap energy of bulk Si in eV and the NC diameter in nm. The size of 2 nm is also consistent with values deduced from TEM measurements of Si-rich SiO_2 layers using very similar annealing procedures [302]. Knowing the size and density of Si NCs allows the calculation of the average distance between neighbouring Si NCs as done in Fig. 5.9 in the upper abscissa.

As expected, the Er-related PL peak at 1.54 μm strongly increases with increasing Si concentration, and only for 15% of excess Si (which is equivalent to a Si cluster density of $\sim 6 \times 10^{19}$ cm^{-3}) the quenching processes introduced by Si become dominant. The PL originating from exciton recombination in Si NCs (open circles in Fig. 5.9) also increases as there is a small fraction of Si clusters which do not couple to Er. However, the Er-related EL (solid squares) continuously decreases and is completely diminished for Si excess concentrations of 10% and above. In contrast to PL, there are at least three additional processes in case of EL which negatively influence either the EL excitation mechanism or the charge carrier transport in SiO_2:

1. In case of PL, an exciton is generated within the Si NC whose recombination on a time scale of μs provides the required energy for exciting the Er^{3+} ions. Contrary to this, mostly electrons are injected during EL which may be trapped by Si clusters but can hardly recombine with a hole. These trapped electrons are not only wasted for EL excitation but also favour the Auger de-excitation of nearby Er^{3+} ions [119]. It was shown that excited Er^{3+} ions can transfer their energy not only to nearby Si NCs but also to nearby traps where the transferred energy is used to release trapped electrons [303]. Whereas at low injection currents, only shallow electron traps are concerned, the problem becomes more severe with increasing injection currents as with increasing applied electric fields, deeper and deeper electron traps can contribute to the Auger de-excitation. Therefore, an efficient exciton *generation* in Si NCs during EL will only be possible by impact excitation of hot electrons.
2. Trapped electrons hamper the efficient excitation because of their Coulomb interaction, too. As shown in the inset of Fig. 5.10, Si clusters have a huge capture cross section for electrons which increases with Si concentration. This means that with increasing charge injection, more and more electrons are trapped in Si NCs (Fig. 5.10) which effectively screen the NCs from other hot electrons by their Coulomb repulsion. As a consequence, more and more Si NCs cannot be excited anymore by hot electrons, and the benefit of an additional pumping of Er by Si NCs is nullified.

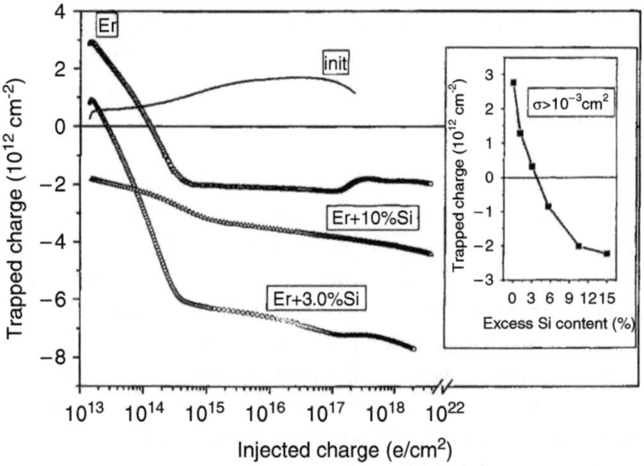

Fig. 5.10 Net trapped charge calculated from the change in applied voltage under constant current injection as a function of the injection charge (after [260]). The *inset* displays the net trapped charge for a capture cross section $> 10^{-13}\,cm^2$ in the early stage of charge injection

3. But most seriously, the energy distribution of hot electrons and the charge carrier transport mechanism are affected. Si NCs are additional scattering centres which lower the average energy of the hot electrons and thus the fraction of hot electrons having enough energy to excite either the Er^{3+} directly or the Si NCs. Moreover, the fraction of electrons whose transport across the SiO_2 layer is mediated by traps or Si NCs is increasing as clearly demonstrated in Fig. 5.11. The FN plots (Sect. 3.1.1) of unimplanted SiO_2 layers and those with a low Si concentration derived from IV data obey the FN formula (2.1) with some deviations caused by charging and contributions due to trap-assisted tunnel injection. So far, the results are consistent with those presented in Chap. 3. However, for Si concentrations of 10% and more, a huge jump in conductivity is observed which can be easily interpreted as the turning point where the trap- or Si cluster-mediated charge transport, most likely by direct tunnelling between clusters, becomes dominant. According to Fig. 5.9, the average distance between Si NCs has shrunken to values below 1 nm for Si concentrations of 10% (which is equivalent to a Si cluster density of $\sim 4 \times 10^{19}\,cm^{-3}$) and more. A similar observation was made for Er-implanted Si-rich oxide layers deposited by PECVD in [145] where the shift of the FN plots with increasing Si concentration towards higher conductivity was accompanied by a drop down of the emitted optical power of up to four (!) orders of magnitude. Recent investigations on the electrical properties of Er in Si-rich SiO_2 report EL power efficiencies in the 10^{-5} range with Poole–Frenkel conduction as the main charge transport mechanism [122].

As shown in Fig. 5.9, the average distance between neighbouring Si clusters is already smaller than the average mean free path of electrons in the CB of SiO_2 of ~ 3 nm for Si concentrations as low as 3% (which is equivalent to a Si cluster

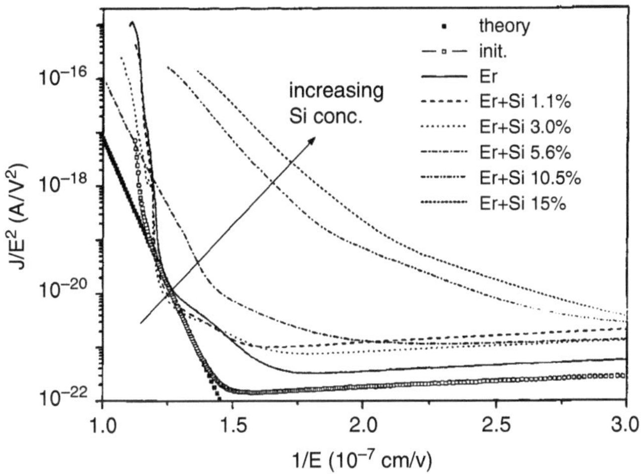

Fig. 5.11 FN plots for Er-implanted MOSLEDs co-doped with Si (after [260]). The change of the charge transport mechanism is clearly seen

density of $10^{19}\,cm^{-3}$). According to Fig. 3.6, an electron released by the Si NC needs an acceleration distance of 10–20 nm in order to get a sufficient high amount of kinetic energy. However, the probability to cover this distance without hitting a Si NC becomes more and more unlikely if the average distance between neighbouring Si clusters is so low.

Finally, it has to be concluded that the suitability of Si NCs for sensitizing Er in case of EL is limited. The problem could be partially solved by constructing MOS devices allowing the simultaneous injection of holes and electrons with comparable intensity as proposed in [119], but up to now such structures were not yet investigated for Er-implanted SiO_2 layers co-doped with Si. Recently, some progress was achieved for Er-doped Si-rich SiO_2 with a very high Si content: in RF magnetron sputtered layers, a maximum in the Er-related EL intensity was observed for a Si:O ratio of 3:2 [304]. Clearly, under such conditions, the charge transport is trap mediated, and the Er^{3+} ions are excited by the recombination of electrons and holes which are injected from the ITO gate electrode and the p-type Si substrate, respectively. Other potential approaches include the use of optical cavities to enhance the Er^{3+} emission [121].

5.3.2 Ge Nanoclusters

5.3.2.1 The Inverse Energy Transfer

After the ambivalent situation with Er coupled to Si NCs, strong increase in case of PL and less optimistic results for EL, the natural question arises how things will change if Si NCs are replaced by Ge NCs. Ge is chemically very similar to

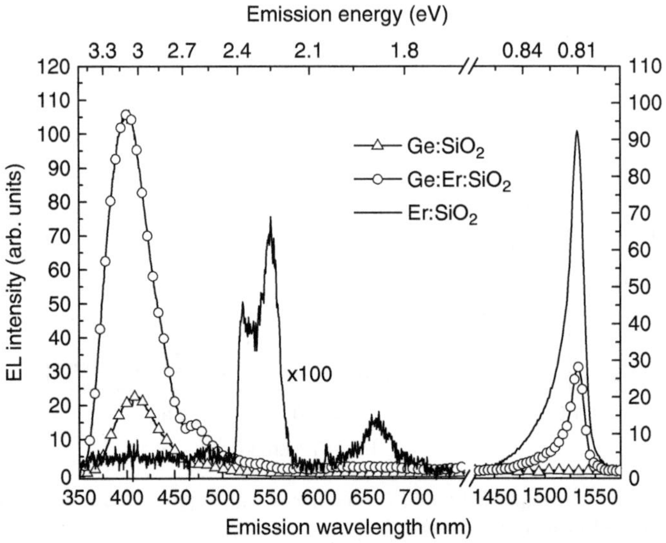

Fig. 5.12 Visible and infrared EL spectra of MOSLEDs implanted with Ge, Er or both elements (after [102])

Si, is a standard material in today's nanoelectronics and is assumed to be superior to the Si counterpart in terms of quantum confinement and inter-band absorption [305]. Generally, there are much less publications dealing with the system Ge and Er [102, 306–309] than with Si and Er. Concerning the Er^{3+} pumping, it was found that photo-excited Ge nanoparticles are not as effective as Si NCs and that the Er^{3+} pumping by amorphous Ge clusters is more effective than that by crystalline ones [307]. However, under certain conditions, the addition of Er to Ge-rich SiO_2 promotes the blue-violet PL [308] or EL [102] from Ge-related ODCs. Finally, it is a matter of fact that the interface between a Ge NC and the surrounding SiO_2 matrix is of much lower quality than the Si–SiO_2 interface.

Figure 5.12 shows the EL spectra of the Ge-rich MOSLEDs with and without Er ions in comparison with MOSLEDs implanted with Er only. The peak concentrations of Ge and Er approximately located in the middle of the SiO_2 layer amount to ~3.5 and 0.3%, respectively, and each implantation was followed by RTA at 1,050°C for either 180 s (Ge) or 6 s (Er). As revealed by TEM (not shown), crystalline Ge clusters with a diameter of about 4 nm are formed both in devices containing only Ge and in those containing Ge and Er. In case of Er-doped SiO_2 layers, ErO_x clusters with a dimension in the range of 2–5 nm were observed [102]. Note that the scales given in the left and right ordinates in Fig. 5.12 are independent of each other as the signals in the visible and IR region were collected by two different detectors.

Regarding the EL spectrum from Er-implanted SiO_2, four distinct peaks around 522, 550, 658 and 1,540 nm can be observed which are assigned to radiative transitions from the $^2H_{11/2}$, $^4S_{3/2}$, $^4F_{9/2}$ and $^4I_{13/2}$ states to the $^4I_{15/2}$ ground state, respectively [233] (an energy level scheme of Er^{3+} can be found in Fig. 5.23, Sect. 5.4.2). The broader EL peak centred at ~400 nm in Ge-implanted SiO_2 is correlated with a $T_1 \rightarrow S_0$ transition within GeODCs [66] which is located either in the bulk of the SiO_2 or in the interface region between the Ge NC and the SiO_2 matrix. Surprisingly, the incorporation of Er *increases* the Ge-related EL intensity at the expense of the Er-related EL intensity, implying an inverse energy transfer from Er to Ge. This is even more astonishing if the fact is considered that the pumped blue-violet emission from Ge is much more energetic than the IR emission from Er^{3+}. However, close inspection of the energy level scheme of Er^{3+} reveals that there are numerous high-lying states which can be excited by hot electrons. In case of Er-implanted SiO_2 layers, these states predominantly relax non-radiatively to low-lying states, and the excited Er^{3+} ions end up mostly in the $^4I_{13/2}$ state. In case of a neighbouring Ge NC with GeODCs on its surface, it seems that there is an efficient energy transfer from the high-lying Er^{3+} levels to the GeODCs. As a result, less Er^{3+} ions relax to the $^4I_{13/2}$ state, causing the decrease of the IR light emission.

In this way, the preliminary model bases on high-lying Er^{3+} levels which have two major decay channels competing with each other: thermalization down to the $^4I_{13/2}$ state and the energy transfer to nearby GeODCs. Unfortunately, the EL decay time can only be measured for the IR Er^{3+} emission and the blue-violet one from Ge as other transitions of interest are too weak to be measured or overlapped by the intense Ge emission. Not surprisingly, the EL decay time of the IR and the blue-violet peak does not change very much. Alternatively, one could suppose that the changes in EL intensity are caused by microstructural changes. However, TEM investigations (not shown) reveal that the introduction of Er in a concentration of 0.3% does not yet change the size and density of the Ge NCs; such changes start for Er concentrations above 0.5% Er [310–312].

Alternatively, someone could argue that the increase of the Ge-related luminescence is due to an increase in the number of GeODC's instead of being a result of the inverse energy transfer. To rule out this, the PL and EL dependency of MOSLEDs with 3.5% Ge on the Er concentration c_{Er} is investigated. In Fig. 5.13a, the PL spectra of these structures excited with a pulsed laser at an excitation wavelength of 266 nm are shown after a delay of 1 μs. This excitation wavelength is found to be non-resonant to the Er^{3+} emission ensuring that only the GeODCs are excited. Whereas the PL intensity continuously decreases with increasing c_{Er}, the PL decay time characteristics monitored at 404 nm exhibit a decay time around 58 μs which is independent of c_{Er} [310]. There are two important conclusions to be drawn, namely, that the number of excited GeODCs *decreases* with increasing c_{Er} and that this is probably not caused by an additional non-radiative decay channel like an energy transfer to Er.

In contrast to this, the blue-violet EL intensity assigned to GeODCs *increases* with increasing c_{Er} up to 0.3% followed by a decrease which is less intense than that observed for PL (Fig. 5.13b). This behaviour can be explained only by assuming an energy transfer process from Er^{3+} to the GeODCs which not only accounts for the observed increase in EL intensity but has to compensate the decreasing number of GeODCs, too.

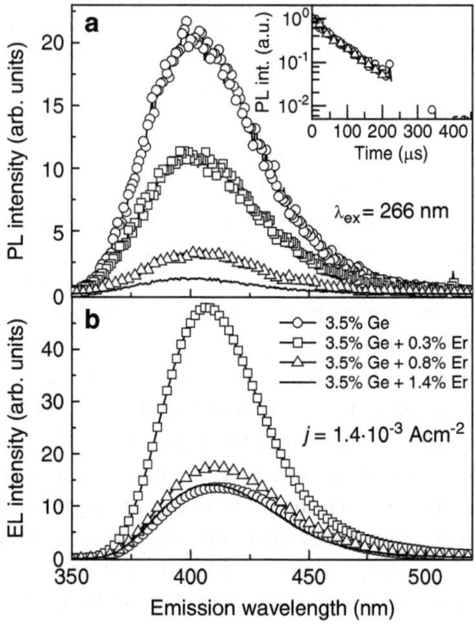

Fig. 5.13 RT PL spectra recorded after a decay time of 1 μs for 3.5% Ge with increasing Er concentration in SiO_2 with an excitation wavelength of 266 nm (**a**). The *inset* shows the corresponding PL decays at a wavelength of 404 nm. Similarly, the corresponding RT EL spectra are shown under constant current injection of $j = 1.4 \times 10^{-3}\,A\,cm^{-2}$ (**b**) (after [310])

5.3.2.2 Er Concentration and Annealing Dependence

The nature of the inverse energy transfer is strongly related to changes in the microstructure, and in order to explore this correlation, a set of devices with varying Ge and Er concentrations was prepared. Ge was implanted in the concentration range of 3.5–11%, followed by FA 950°C 60 min, followed by Er implantation in the range of 0.3–1.4%, and finished with FA 800–1,100°C 30 min. In the following, the dependence of the microstructure on c_{Er} and the final annealing step as well as its impact on the energy transfer process is discussed. For further details, see the corresponding literature [311, 312].

Figure 5.14 displays the EL spectra of MOSLEDs annealed with FA 900°C and containing 7.4% Ge plus different Er concentrations (a) as well as for devices with different annealing temperatures T_a and with 7.4% Ge + 0.5% Er (b). The increase in the blue-violet EL intensity in Fig. 5.14a with increasing c_{Er} is very similar to what is shown in Fig. 5.13; the maximum EL intensity is reached for 0.5% in this case. However, both the IR EL due to Er^{3+} and the Ge-related EL decrease continuously with increasing T_a which is interpreted as a decrease of the number of the corresponding LCs. Indeed, RBS and TEM investigations (not shown) reveal that there is a strong diffusion of Ge towards the two interfaces of the SiO_2 layer and that the Ge NCs grow in size with increasing T_a. Similar to the discussion in Sect. 4.4 about RE diffusion, RE clustering and their impact on EL, both processes decrease the number of potential LCs. In contrast to Ge, the Er distribution remains unchanged with increasing T_a. Further details about the microstructural development of Ge in SiO_2 can be found elsewhere [67, 183].

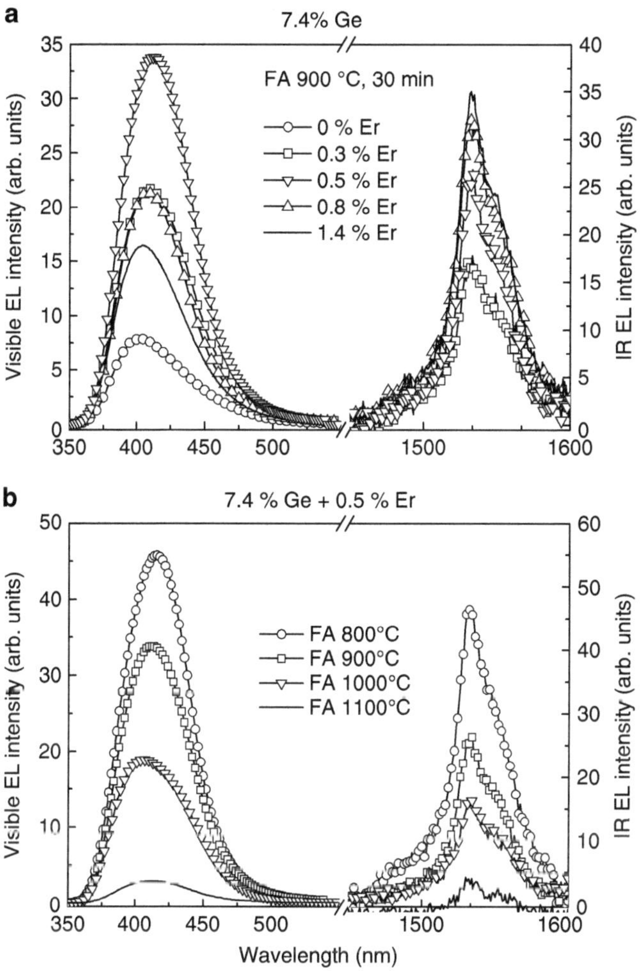

Fig. 5.14 The visible and infrared EL spectra recorded for MOSLEDs containing 7.4% Ge plus different Er concentrations (**a**) and with increasing annealing temperature for 7.4% Ge and 0.5% Er (**b**) [311]

Figure 5.15 displays the development of the μ-Raman spectra of Ge-rich SiO_2 layers with Ge concentration (a), Er concentration (b) and the annealing temperature (c). The spectra are dominated by the characteristic feature of the Ge–Ge mode of crystalline Ge [313,314] whose position and FWHM amounts to 300.2 cm^{-1} and 3.1 cm^{-1} in case of bulk material, respectively. Without additional Er, the Ge peak for 7.4% Ge is located at 297.2 cm^{-1} with a FWHM of ~6 cm^{-1} (Fig. 5.15a) which is a consequence of strain and phonon confinement in NCs [313]. With increasing c_{Er} up to 0.5%, the Ge peak becomes more intense and sharper down to a FWHM ~5.2 cm^{-1}, followed by a reduction in intensity and a broadening for a higher c_{Er}.

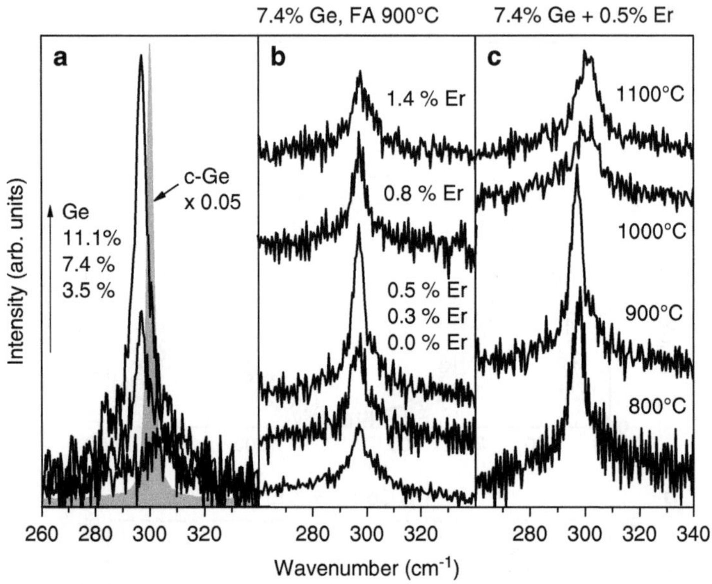

Fig. 5.15 The μ-Raman spectra of c-Ge (*grey area under the curve*) and SiO_2 layers containing 3.5, 7.4 and 11.1% Ge (**a**); the c-Ge data are divided by 20 for clarity. The dependency of the μ-Raman spectra on Er concentration (**b**) and annealing temperature (**c**) displays a sharp Ge–Ge Raman peak for 0.5% Er and FA 900°C, respectively (after [311])

The corresponding TEM investigations (not shown, but similar to Fig. 5.17) reveal that the Ge NC size remains nearly unchanged up to 0.5% Er, while it decreases in size for $c_{Er} > 0.5\%$ due to the sputtering and fragmentation of bigger Ge NCs during Er implantation. According to the phonon confinement in Ge NCs, a broadening and an intensity reduction of the Ge–Ge Raman peak are expected with decreasing NC size which fits to the experimental results for $c_{Er} > 0.5\%$. In order to explain the initial sharpening of the peak with increasing c_{Er}, it can be assumed that a part of the implanted Er ions accumulates in the vicinity of Ge NCs, forming a shell-like structure. Such a structure with Er surrounding a NC was recently detected in case of silica layers implanted with Si and Er [315]. In this framework, the phonon wave function is more or less completely confined within Ge NCs before Er doping, whereas the Er ions residing close to Ge NCs may allow a partial propagation of the phonon wave function beyond the NC boundary for 0.3 and 0.5% Er by declining the relaxation condition of the phonon selection rules [316].

The Ge–Ge Raman peak intensity increases with T_a up to 900°C, while it drops drastically and moves towards the crystalline Ge peak position (Fig. 5.15c) for $T_a > 900$°C. The T_a dependency of the Ge–Ge mode is the result of the following effects: (1) the increase in Ge NC size with increasing T_a sharpens the peak while (2) the diffusion of Ge towards the interfaces followed by the removal of the Er shell during Er phase formation (see next chapter) suppresses the peak intensity with a subsequent increase in FWHM. It is clear from Fig. 5.15c that the second effect dominates for $T_a > 900°C$. For further details, especially the slight shift of the Ge–Ge mode peak position, see [311].

The microstructural development of the Ge NCs was also traced by XRD measurements. In Fig. 5.16, the formation of Ge NCs is indicated by the appearance of the Ge(111) and (220) peaks which exhibit a similar Er dependence as the Ge–Ge Raman mode: a sharpening and an increase in intensity for Er concentrations up to 0.5% followed by broadening and intensity reduction for $c_{Er} > 0.5\%$. Basically, the same explanation as for the Raman mode applies, although this case is more complex due to the various influences on peak broadening [317]. The average Ge NC diameter d can be estimated by using the Scherrer formula [314]:

$$d = \frac{0.89\lambda}{\beta(2\theta_B) \cdot \cos\theta_B}, \tag{5.12}$$

with λ and $\beta(2\theta_B)$ being the X-ray wavelength (0.154 nm) and the FWHM of the concerned peak at the Bragg angle θ_B. Using this formula, the average Ge NC size is estimated to be ~8.1 nm for Er concentrations up to 0.5% and to decrease to ~6.4 nm for $c_{Er} > 0.5\%$. For 0.8% of Er, nanocrystallites of Er_2O_3 represented by the (222) and (440) peaks of the hexagonal phase appear which, however, completely disappear at 1.4% Er. This phenomenon suggests that above 0.8% Er, the Er_2O_3 nanocrystallites are either transformed into an amorphous phase or form a non-crystalline composite structure like Er-pyrogermanate [318], depending on the

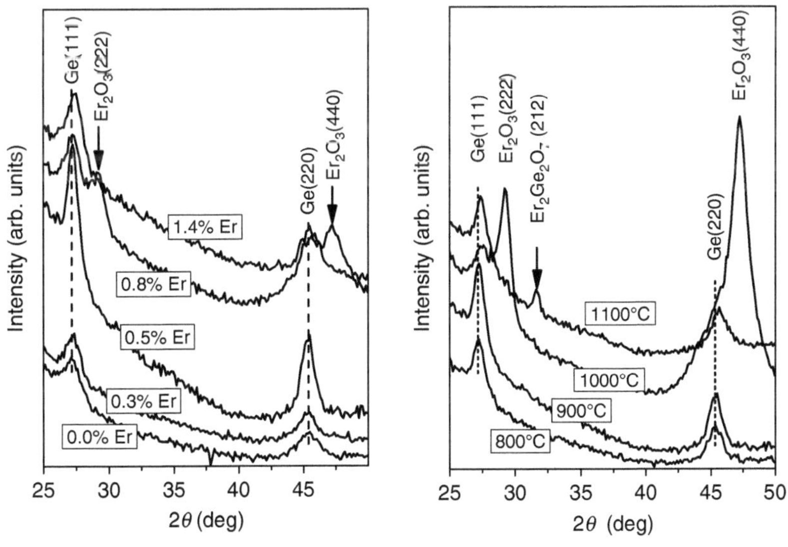

Fig. 5.16 XRD patterns with increasing c_{Er} for 7.4% Ge annealed at 900°C (**a**) and with increasing T_a for 7.4% Ge and 0.5% Er (**b**) [311]

Er-to-O ratio and the interaction strength in the Ge-rich environment. Interestingly, while the Er oxide precipitates are amorphous only in the Er-doped SiO_2 layers [120], the Er_2O_3 clusters are now crystalline in the presence of Ge NCs for 0.8% Er (Fig. 5.16a).

The interesting question what happens to the crystalline Er_2O_3 precipitates after dissolution is answered in Fig. 5.16b showing the XRD pattern of MOSLEDs with 7.4% Ge and 0.5% Er as a function of T_a. Similar to the concentration dependence, the (222) and (440) peaks of the hexagonal Er_2O_3 phase appear at 1,000°C and disappear for higher temperatures. Instead of this, the (212) reflection from $Er_2Ge_2O_7$ nanocrystallites emerges at 1,100°C, whose tetragonal structure is known to be non-centrosymmetric and shows anisotropic magnetic properties [318].

5.3.2.3 Microstructural Development of the Ge–Er System

Finally, the following scenario for the microstructural development in Er-implanted MOSLEDs co-doped with Ge and its impact on the inverse energy transfer can be deduced. After Ge implantation followed by annealing spherical, randomly oriented Ge NCs are present (Fig. 5.17a, e), and a fraction (~10%) of Ge is diffused

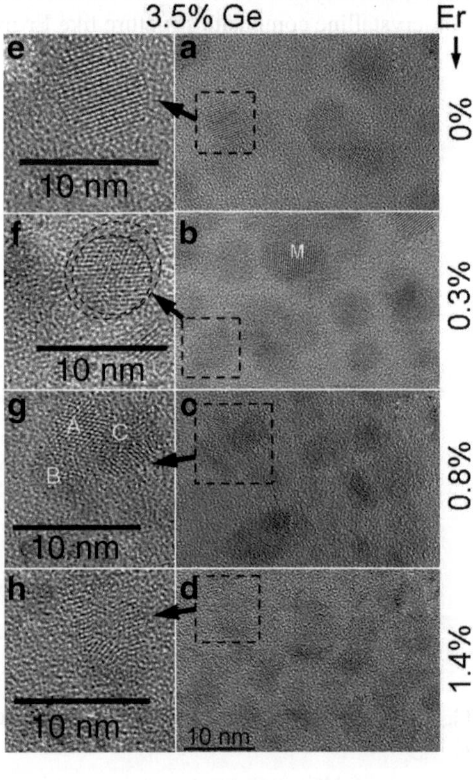

Fig. 5.17 Cross-sectional TEM images of the only Ge-rich SiO_2 (**a**) and Er-doped Ge-rich SiO_2 with concentrations of 0.3% (**b**), 0.8% (**c**) and 1.4% (**d**) Er for 3.5% Ge. The corresponding HRTEM images are shown in (**e, f**), showing the gradual reduction of the average NC size from ~6.1 ± 2 nm (**a**), ~5.9 ± 2 nm (**b**), ~3.5 ± 2 nm (**c**) and ~2.7 ± 1 nm (**d**), respectively [312]

towards the $Si–SiO_2$ interface during annealing as revealed from RBS. Generally, there is only a little difference between 3.5 and 7.4% of Ge; Ge cluster sizes and the optimum Er concentration are somewhat larger in the latter case.

After low-dose Er implantation (0.3%), hardly any change was observed in the overall microstructure except the appearance of a few bigger NCs with diameters of $\sim$10 nm (marked by "M" in Fig. 5.17b). A closer inspection of a NC reveals the formation of a shell-like structure where a thin amorphous layer (light grey contrast) is found situated at the $NC–SiO_2$ interface which is indicated by dashed circles in the adjacent HRTEM image (Fig. 5.17f). The shell-like structure resembles an Er-rich zone around a Si NC in [315], but in addition, the ion beam-induced amorphization of Ge NCs [319] has to be considered. This Er shell, whose existence is also indicated by μ-Raman, XRD and TEM investigations, is assumed to be the optimum condition for an efficient inverse energy transfer, and consequently 0.3–0.5% of additional Er gives the most intense Ge-related EL (Figs. 5.13 and 5.14).

An increase of the Er dose leads to fragmentation and a partial amorphization of the Ge NCs (Fig. 5.17c) and triggers the formation of crystalline Er_2O_3 precipitates, followed by a phase of dissolution and the formation of $Er_2Ge_2O_7$ nanocrystallites. It has to be considered that the RBS profile of Er does not change very much with temperature, implying that the phase formations of Er are the result of local rearrangements rather than that of Er diffusion. Indeed, for 3.5% of Ge, the average size of Ge NCs was found to decrease gradually from $\sim$6.1 to 2.7 nm with increasing c_{Er} from 0 to 1.4%, respectively. Since the second annealing was carried out at 900°C, which is lower than the melting temperature of Ge NCs [320], it is unlikely to have an amorphous-to-crystalline transition of the damaged Ge clusters. The corresponding HRTEM image (Fig. 5.17g) depicts three distinct regions, indicated by "A" (crystalline), "B" (amorphous in light grey) and "C" (amorphous in dark grey). In fact, the observed change in contrast in B and C can be classified in view of the formation of a variety of non-crystalline Er composites, which depend strongly on the ratio of Er and O atoms and the interaction strength in a Ge-rich environment [311]. The Er ion beam-induced damage becomes more pronounced at 1.4% Er (Fig. 5.17d) where the HRTEM image of severely damaged NCs is projected in Fig. 5.17h. Whereas the Er implantation decreases continuously the number of GeODCs (Fig. 5.13), the removal of the Er shell and the incorporation of Er into Er composites strongly reduce the intensity of the inverse energy transfer for Er concentrations of 0.8% and above. The behaviour of the IR Er-related EL (Fig. 5.14) is the result of two opposite dependencies: on one hand, higher implantation doses increase the number of Er^{3+}, but on the other hand, an increasing fraction of Er either takes part in the inverse energy transfer process (low concentration) or is incorporated in Er composites (higher concentration). Further details can be found in [102, 120, 310–312].

5.3.3 Si and Ge Nanoclusters: A Short Comparison

After all these details, this small chapter tries to work out the main differences between Si and Ge NCs leading to such a different behaviour, especially for EL. For the following discussion, the three different LCs, namely the NC (either Si or Ge), the closely located Er^{3+} ion and an ODC located in the surface region of the NC (either Si or Ge related), and their relationships among each other have to be considered. Figure 5.18 illustrates the situation showing the three LCs with their ground and the excited states of interest in a basic scheme. Possible radiative and non-radiative transitions are indicated by solid and dashed arrows, respectively. The PL technique has the possibility to exclude a part of these transitions by selective excitation, but in case of EL all the three species are excited simultaneously. In addition, the EL excitation cross section for Er^{3+} is higher than that in case of PL as the optical selection rules do not apply.

Starting the discussion with Si, it can be assumed that the interface quality between the Si NC and the SiO_2 matrix is good, resulting in a fairly long exciton lifetime within the NC which increases the probability for an energy transfer NC $\rightarrow$ Er. The corresponding transition rate $k(\text{NC} \rightarrow \text{Er})$ does not need to be high; it is sufficient to be larger than the transition rate for an exciton recombination without an energy transfer. Because of the high interface quality, the number of defects is low, especially those of Si-related ODCs. Consequently, Er excited into high-lying states will predominantly thermalize to the $^4I_{13/2}$ state as the number of Er^{3+} ions is high compared to the number of Si NCs and Si-related ODCs. In addition, luminescence decay times of Si-related ODCs are usually long with values in the ms range [241] which favours a possible energy transfer ODC $\rightarrow$ Er. Therefore, the excited Er^{3+} ion does not face a substantial "energy loss" to other LCs and can even be pumped by Si NCs and Si-related ODCs, although this effect is weaker compared to PL because of the high EL excitation cross section of Er^{3+}.

The situation changes completely with Ge. The interface between the Ge NC and the SiO_2 matrix is of lower quality compared to the case of Si, which causes two consequences: At first, the exciton lifetime within a Ge NC is shortened, and an energy transfer NC $\rightarrow$ Er is less likely, even if $k(\text{NC} \rightarrow \text{Er})$ is the same as in case

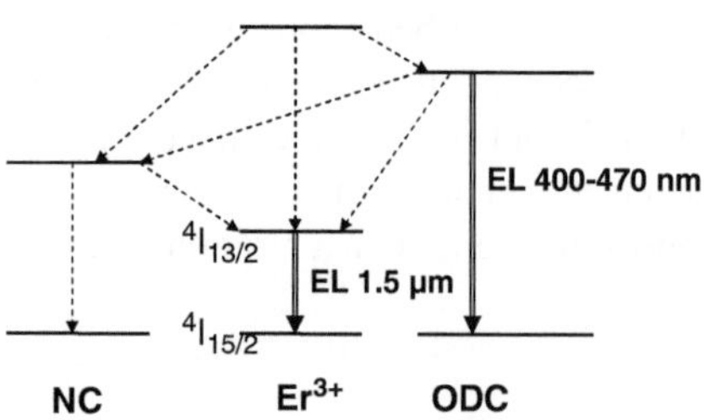

Fig. 5.18 Basic energy level scheme of a Si or Ge NC, an Er^{3+} ion and a Si- or Ge-related ODC. Possible radiative and non-radiative transitions are indicated by *solid* and *dashed arrows*, respectively

of Si. Second, the number of defects, especially of GeODCs, is higher. Therefore, a substantial part of highly excited Er^{3+} ions transfer their energy to GeODCs instead of relaxing down to the $^4I_{13/2}$ level. Moreover, the EL decay time of GeODCs is with 100 μs or less [100] significantly lower than that for Si-related ODCs which limits the possible energy transfer ODC → Er. As a result, Er^{3+} can take little advantage of the presence of Ge NCs, and a larger part of the Er^{3+} ions is even forced to transfer their energy to nearby GeODCs.

The presented model is qualitative but consistently explains the small ability of Ge NCs to act as sensitizers for Er^{3+} and the inverse energy transfer from Er^{3+} to the GeODCs. Whereas the first point is quite obvious, the second one, however, requires an additional assumption, namely that the transition rate $k(\mathrm{Er} \rightarrow \mathrm{GeOCD})$ is sufficiently high in order to outnumber the internal relaxation of Er^{3+} and to cause the observed enhancement of the Ge-related EL described in the previous chapters.

5.4 Pumping by Other Rare Earth Elements

Pumping by other RE^{3+} ions implies a resonant energy transfer from a *pumping* RE element to a *pumped* RE element and may occur if the energy of a transition within an excited, pumping RE^{3+} ion fits the absorption energy of an RE^{3+} ion to be pumped and being in the ground state. $4f$ intrashell transitions of the pumping RE^{3+} ion have the advantage of being optically forbidden and thus having a long EL decay time. Therefore, the corresponding intrinsic decay rate ($k_1 + k_2$ in Fig. 5.2) is low and already moderate transition rates (k_3) for the energy transfer to the pumped RE^{3+} ion can have a significant effect. In the following, two examples are given which demonstrate that this type of pumping can be quite efficient under optimum conditions but also can fail if the prevailing circumstances are less favourable.

5.4.1 Pumping of Cerium by Gadolinium

Figure 5.19a shows the EL spectrum of 100 nm thick MOSLEDs implanted with 1% Ce and co-implanted with Gd with varying concentration. The implantation energies were chosen in such a way that both implantation profiles peak in the middle of the oxide lay in order to ensure a maximum overlap between them. As clearly seen, the EL intensity of the blue EL band around 440 nm ascribed to a $5d \rightarrow 4f$ transition of Ce^{3+} (Sect. 4.1.4.1) gets stronger with increasing Gd concentration at the expense of the Gd-related EL at 316 nm. In case of 1.5% Gd (not shown), the additional co-doping with 1% Ce leads to a quenching of the Gd-related EL by more than one order of magnitude [130]. Comparing the MOSLED with only 1% Ce with that having 1% Ce and 3% Gd, the EL intensity increases by a factor of 4 which corresponds to an increase of the excitation cross section from $4.8 \times 10^{-13}\,\mathrm{cm}^2$ to

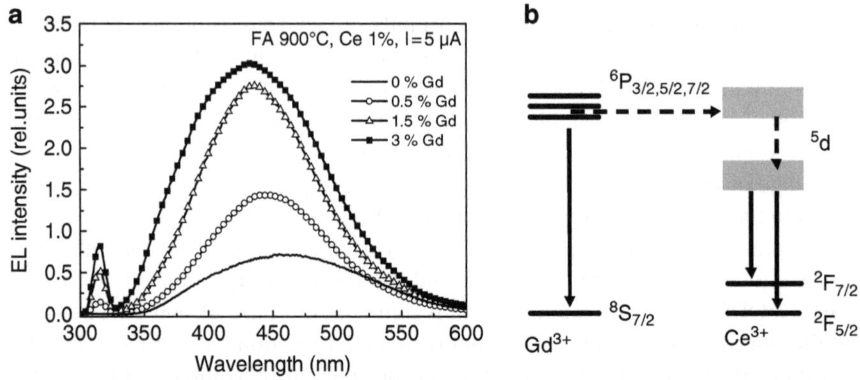

Fig. 5.19 EL spectrum of 100 nm thick MOSLEDs implanted with 1% Ce and co-implanted with Gd with varying concentration (**a**). All devices were annealed by FA 900°C [130]. The proposed energy transfer process is schematically shown on the *right* (**b**)

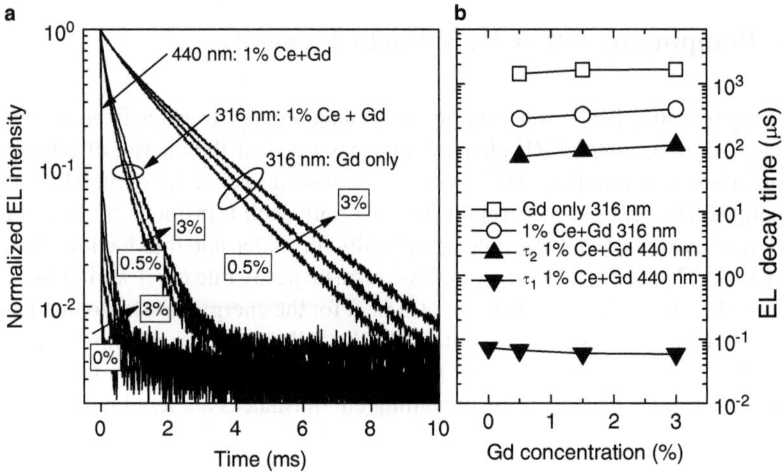

Fig. 5.20 Normalized EL decay time characteristics of the blue Ce-related EL band and the UV Gd-related EL line for MOSLEDs implanted with Ce, with Gd, and Ce co-implanted with Gd (**a**). The observed decay times and their dependence on Gd concentration are given in (**b**) (after [130])

3.5×10^{-12} cm^2 [130]. The excitation cross section in that work was derived from *IV* measurements according to (4.4).

The observed behaviour can be explained by assuming an energy transfer from the excited Gd^{3+} ion to Ce (Fig. 5.19b) which can easily be traced by measuring the EL decay times as displayed in Fig. 5.20. For MOSLEDs implanted with Gd, only an EL decay time of the 316 nm line between 1.43 and 1.66 ms is measured which drops down to values between 280 and 400 μs as soon as Ce is co-implanted. According to Sect. 5.1.3, this is the expected behaviour if the resonant energy transfer from Gd to Ce is directly competing with the intrinsic decay of the Gd^{3+} ion. The transition rate of the energy transfer adds to the intrinsic decay rate of the $^6P_{7/2}$ state of Gd^{3+} which increases the total relaxation rate of this state and thus decreases its decay time. The decay of the Ce-related luminescence is composed of a fast and slow component with decay constants τ_1 and τ_2, respectively. Whereas in case of MOSLEDs implanted only by Ce a decay constant of $\tau_1 = 75$ ns is measured, the co-implantation with Gd reduces τ_1 to 60 ns together with the appearance of the slow component with $\tau_2 = 70$–110 μs. The latter one is interpreted as the slow excitation of Ce^{3+} via the energy transfer from Gd^{3+} leading to a delayed population of the excited Ce^{3+} state, for which reason τ_2 is assumed to be characteristic for the energy transfer process from Gd to Ce. Peak positions and the fast component of the EL decay of the blue Ce^{3+} peak at 440 nm are similar to the behaviour of Gd_2SiO_5:Ce scintillators excited by UV light or γ rays [321]. It should be still mentioned that all three groups of decay time characteristics have the same interesting signature: a slight increase of the EL decay time with increasing Gd concentration.

Finally, the resonant energy transfer was also studied by comparing the PL excitation spectrum of Ce^{3+} and the PL emission spectrum of Gd^{3+} as shown in Fig. 5.21. The PL excitation spectrum of Ce^{3+} is characterized by a maximum at 300–310 nm with two satellites at 275 and 340 nm, which overlap well with the $^6P_{7/2} \rightarrow {}^8S_{7/2}$ transition of Gd^{3+}. This overlap together with the dependency of the EL intensity and EL decay time on the Gd concentration gives convincing evidence for the resonant energy transfer from Gd to Ce.

5.4.2 *Pumping of Erbium by Gadolinium*

After the encouraging results for the energy transfer of Gd to Ce, the question arose whether other RE combinations work as well or not. In Fig. 5.22, the EL spectra of Er-implanted MOSLEDs co-implanted with Gd to different concentrations and annealed with FA 900°C are shown. For devices with only 2% Er, weak EL in the visible spectral region and a strong EL peak at 1,540 nm are observed. Whereas the green EL lines at 530 and 550 nm can be assigned to radiative transitions from

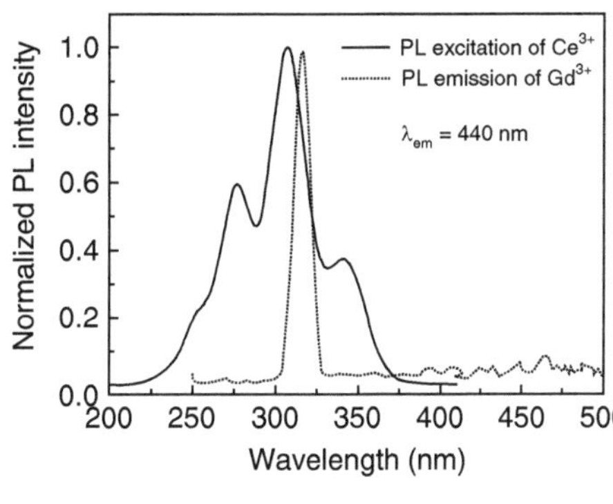

Fig. 5.21 PL excitation spectrum of Ce^{3+} monitored at 440 nm and the PL emission spectrum of Gd^{3+} for MOSLEDs with 1% Ce and 1.5% Gd (after [130])

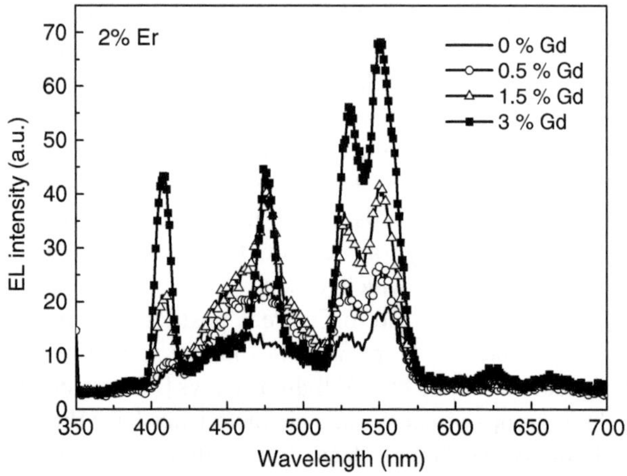

Fig. 5.22 EL spectrum of MOSLEDs with a 100 nm thick SiO_2 layer implanted with 2% Er and co-implanted with Gd in different concentrations. All devices were annealed by FA 900°C (after [334])

the $^2H_{11/2}$ and the $^4S_{3/2}$ state to the $^4I_{15/2}$ ground state, respectively, the weak blue-violet EL peaks can be either due to radiative transitions of higher lying Er^{3+} levels or due to implantation-induced defects in the SiO_2 layer. The EL of these implantation-induced defects is dominated by a broad emission at 460 nm which is characteristic for ODCs.

The incorporation of additional Gd leads to an increase of the Er-related EL in the *visible* range but does not have a significant impact on the EL intensity at 1,540 nm. This behaviour can be explained by considering the energy level schemes of Gd^{3+} and Er^{3+} as depicted in Fig. 5.23. According to this, the energy transfer from Gd to Er starts with the non-radiative transition from the $^6P_{3/2}$ state of Gd to the $^8S_{7/2}$ ground state. The released energy is used to excite Er^{3+} ions being in the $^4I_{15/2}$ ground state to higher lying states, most probably to the $^2K_{13/2}$ or the $^2P_{3/2}$ state. Investigations devoted to the upconversion PL in cubic Gd_2O_3:Er [322] and Er^{3+} doped YAG crystals [323] revealed that the $^2P_{3/2}$ state can radiatively decay to the $^4I_{11/2}$ or $^4I_{13/2}$ state by emitting a photon at 470–480 nm and around 410 nm, respectively. However, this decay strongly increases the population of the $^4I_{13/2}$ state which would cause an adequate increase of the IR EL at 1,540 nm. As this is not observed, it is more likely that the $^2P_{3/2}$ state decays non-radiatively to lower lying states which in turn contribute to the visible EL by radiative transitions to the $^4I_{15/2}$ ground state. Therefore, the blue-violet peaks at 408 and 475 nm are most probably due to radiative transitions from the $^2H_{9/2}$ and $^4F_{5/2}$ states, respectively. This scenario is supported by the investigation of photon avalanche processes in Er^{3+} excited to high energy states [324] in which a strong green emission originating from the $^4S_{3/2} \rightarrow {}^4I_{15/2}$ transition in Er^{3+} was found. Furthermore, the increase of the Er-related EL in the visible range is accompanied by an adequate decrease of the Gd-related EL (not shown).

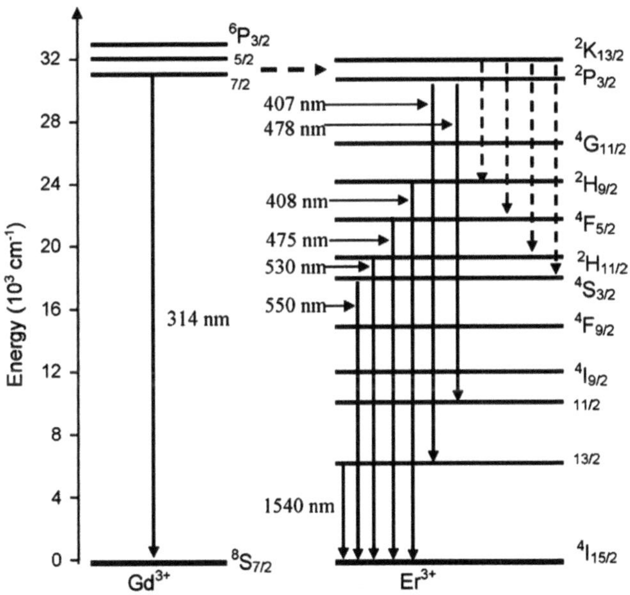

Fig. 5.23 Energy level scheme of Gd^{3+} and Er^{3+} ions. *Solid* and *dashed arrows* denote radiative and non-radiative transitions (after [334])

5.5 Fluorine Co-Doping

The use of fluorine co-doping is motivated by the beneficial effect of fluorine on the defect formation and the optical properties of SiO_2. It was found that the incorporation of fluorine considerably decreases the number of E' centres, NBOHCs and strained Si–O–Si bonds originating from three- and four-membered ring structures [325]. As a result, the authors observed an increase of the UV transparency of a 5 mm thick SiO_2 layer at 157 nm from 10% up to 80%. Further investigations revealed that fluorination decreases the number of oxide charges and Si–SiO_2 interface defects [326] during the thermal growth of SiO_2 on Si. This beneficial effect is explained by the termination of dangling Si bonds and the replacement of Si–H bonds with a binding energy of $E_b = 3.18$ eV by Si–F bonds ($E_b = 5.73$ eV) [327, 328], which in turn leads to a reduction of trapping and hopping sites and thus to a decrease of leakage currents by up to two orders of magnitude [329]. Generally, an increase of silylene-like centres, namely, Si–F and Si–O–F bonds, is expected after fluorine implantation followed by annealing [138].

Figure 5.24a displays the EL spectra of MOSLEDs implanted with 2% Gd, 3% F or 2% Gd followed by an implantation with 3% F. In comparison to the MOSLED with only Gd, the incorporation of fluorine increases the EL intensity of the Gd-related EL peak at 316 nm almost by a factor of 2 with a concurrent narrowing of the FWIM of the 316 nm line from ~160 to ~110 meV. The visible part of the EL spectra is caused by defect luminescence. Whereas for the unimplanted devices the weak

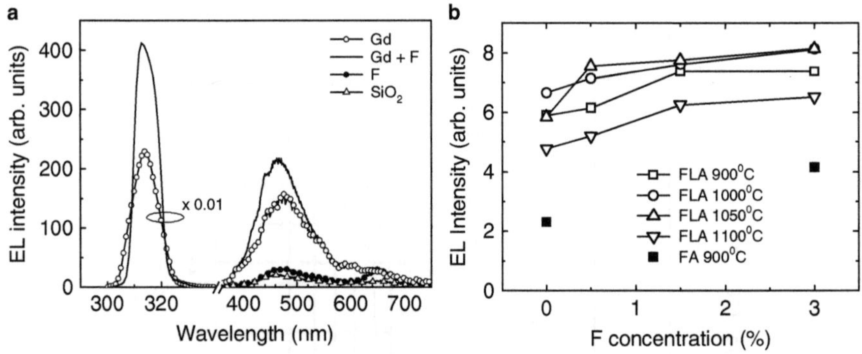

Fig. 5.24 EL spectra of MOSLEDs implanted with 2% Gd, 3% F and subsequently 2% Gd and 3% F (**a**). The maximum EL intensity at 316 nm as a function of fluorine concentration and for different annealing conditions (**b**) (after [138])

blue and red EL at 460 and 650 nm is due to ODC and NBOHC centres, respectively, the fluorine co-implantation slightly increases this EL. In detail, silylene-like centres can contribute to the blue EL, whereas peroxy defects created during fluorine implantation are assumed to enhance the red EL [166]. As the assignment of the defect EL bands to a specific defect is difficult in case of Gd + F co-implantation, it cannot be decided from the present data whether the beneficial role of fluorine in reducing defects is outbalanced by the creation of new implantation-induced defects or not. Indeed, an optimum fluorine dose around $10^{15}\,F\,cm^{-2}$ was found to create a minimum of traps at the Si–SiO_2 interface with higher doses reducing the advantage of fluorine [330]. In the present case, 3% F is equal to an implantation dose of $1.3 \times 10^{16}\,F\,cm^{-2}$. Figure 5.24b summarizes the effect of fluorine co-implantation and different annealing conditions on the Gd-related EL intensity. Whereas for FA 900°C the enhancement amounts to a factor of almost 2, for MOSLEDs annealed by FLA – which already show an enhanced UV emission compared to FA – the EL intensity increases by about 40%.

There is a discussion in literature that fluorine can enhance the number of optically active RE^{3+} centres via RE–F_3 creation [331] and that excited silylene-like centres can transfer their energy to nearby RE^{3+} ions [332,333]. However, the main impact of fluorine is most probably the aforementioned reduction of defects leading to a reduced scattering rate for hot electrons. As a consequence, they have a somewhat higher kinetic energy under the same electric field allowing a larger fraction of electrons to excite RE^{3+} centres. Certainly, this implies that the effect of defect reduction by fluorine is larger than that of defect creation by implantation. Another speculation is that the RE–F_3 creation increases the number of optically active RE ions by avoiding or decelerating the clustering of RE atoms during annealing.

Chapter 6
Stability and Degradation

This chapter covers two aspects of degradation, namely the identification of the main degradation mechanisms and their description, and strategies to avoid or to slow down such processes. The degradation of Si-based light emitters is composed of two components: the electrical degradation of the gate oxide, also called the "wear-out" of the MOS structure, and the EL quenching with increasing charge injection. Consequently, Sects. 6.1 and 6.2 treat these two aspects of degradation, followed by a discussion about the important role of a SiON buffer layer for the stability enhancement (Sect. 6.3) and the usefulness of a potassium co-implantation (Sect. 6.4).

6.1 Electrical Degradation

6.1.1 Wear-Out Mechanisms in MOS Structures

Wear-out phenomena are degradation processes in which defects are generated during the electrical operation of a device, and which lead finally to its breakdown (BD). Generally, these processes can be divided into the phase of defect generation, the phase of defect accumulation or coalescence and the final BD in which a conductive path is created across the SiO_2 layer. The BD can be characterized by either the BD field E_{BD} or the time-to-breakdown t_{BD} under constant voltage or constant current injection, respectively. E_{BD} is calculated from *IV* measurements by using the voltage at which a current surge marks the BD. In case of constant current injection, this current surge is avoided, and the BD is now characterized by a sudden drop down of the applied voltage. However, for very thin SiO_2 layers, the phenomenon of a progressive BD is known in which the voltage or the current changes not abruptly, but in an erratic, fluctuating way. The charge-to-breakdown Q_{BD} is a more feasible quantity than t_{BD}, especially when comparing different injection current levels, and can be easily determined by integrating the injection current over time. Most models comply with the idea that the BD occurs if a critical defect density N_{BD} is exceeded and that Q_{BD} and N_{BD} obey in the first approximation the equation

$$Q_{BD} = \frac{qN_{BD}}{P_G}, \tag{6.1}$$

with P_G as the defect generation probability per injected electron and area unit [335].

The different degradation mechanisms differ by the way of creating defects, the type of defects and the dependencies of the BD parameters on electrical operation conditions. The first category of degradation mechanisms comprises thermochemical models in which the pure presence of a strong electric field, eventually in combination with an elevated temperature, is enough to induce a bond breaking which especially applies to distorted or weak Si–H or Si–Si bonds in ODCs [336, 337]. A detailed model for Si–O bonds is also given in [338]. However, whereas this model was more popular in the 1980s, present models favour the energy release of hot electrons in combination with charge trapping to be the most relevant degradation mechanism. As discussed in Sect. 3.1.2, electrons are accelerated in the CB of SiO_2 and reach an equilibrium distribution of their kinetic energy with an average energy of a few eV and a high energy tail with energies exceeding the bandgap energy of SiO_2. If released in inelastic scattering events, this energy can easily create new defects, either directly by bond breaking or indirectly by band-to-band or trap-to-band impact ionization (Sect. 3.4.2). In fact, in studies investigating the trapping characteristics of SiO_2 layers as a function of the applied electric field [204], it was found that trap generation under charge injection starts if the average energy of hot electrons exceeds a value of 2.3 eV (Fig. 6.1). As this value is already reached for electric fields of 4 MV cm^{-1} (Fig. 3.5), but current injection at suitable levels is not achieved for local fields below 8 MV cm^{-1}, this type of degradation is of relevance for the further discussion of RE-implanted MOSLEDs.

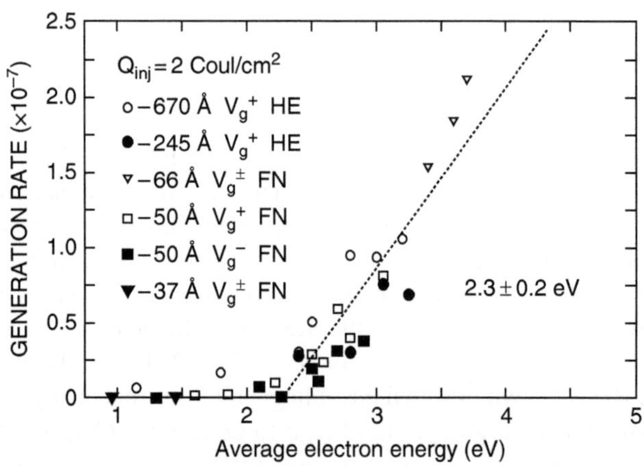

Fig. 6.1 Trap generation rate in traps per injected electron as a function of the average energy of the hot electrons for SiO_2 layers of different thicknesses (after [204])

However, defect and trap generation is not homogeneously distributed across the SiO_2 layer. Because of the initial phase of acceleration, there is a "dark space" behind the injecting interface in which electrons have not enough energy to induce impact ionization or other defect creation processes [208]. This dark space is very similar in its extension to the dark zone discussed in Sect. 5.2.2. In contrast to this, an enhanced defect generation rate is expected at the opposite interface of the SiO_2 layer, as (1) the high energy tail of the equilibrium energy distribution of the electrons is most pronounced there and (2) an electron can release its kinetic *plus* the band offset energy between the CB of SiO_2 and the anode material (the system Si–SiO_2–SiON–ITO is discussed in Sect. 6.3). Besides impact ionization, the energy release at the anode interface gives rise to two other but similar BD mechanisms, namely, to the anode hole injection and the anode hydrogen release, which were already discussed in Sect. 3.4.2. Both processes generate positive charge carriers which migrate towards the Si–SiO_2 interface and can create defects on their way by reacting with defect precursors in the bulk and at the Si–SiO_2 interface [222, 339, 340]. The dependence of the defect creation efficiency on the electron energy at the anode can be found in [341].

Generally, Q_{BD} values and thus the stability of the devices is known to decrease with increasing operation temperature [342], with increasing device area [343] and increasing oxide thickness [209, 210]. The latter point applies mostly to ultrathin SiO_2 layers where the maximum energy an electron can gain during its passage through the oxide layer is strongly reduced. However, the superiority of ultrathin SiO_2 layers compared to thick SiO_2 layers is strongly field-dependent [344] and may even turn into the opposite for special conditions [345]. Furthermore, the introduction of hydrogen also causes a decrease of the stability either by a frequent passivation and depassivation of P_B centres at the Si–SiO_2 interface [346, 347] or by bond breaking of strained Si–O or Si–Si bonds in the bulk followed by hydrogen termination of the dangling bonds [340, 348].

6.1.2 Statistical Description of the Breakdown

The BD of a MOS capacitor is a very local event which occurs on length scales of 10–100 nm [349] up to a few micrometres [196]. Compared to this, the area of a MOSLED is large and can be imagined as a composition of a large number of such small BD areas. Clearly, the whole device breaks down if the first and weakest local spot is breaking down, for which reason the BD has a so-called "weakest link character". Such a behaviour is usually described by a Weibull distribution in which the cumulative failure rate F is given by

$$F = 1 - \exp[-\exp(K)] \quad \text{and} \quad K = \frac{\ln(t) - \ln(t_m)}{\sigma_D}, \tag{6.2}$$

with σ_D being a dispersion parameter characterizing the width of the distribution [196]. t_m leading to $K = 0$ is the modal lifetime after which 63.2% of the devices of a sample have been broken down. For practical purposes, (6.2) is transformed into

$$\ln\left[\ln\left[(1-F)^{-1}\right]\right] = \frac{\ln(t) - \ln(t_m)}{\sigma_D} \tag{6.3}$$

and the left side of the equation is plotted vs. $\ln(t)$, resulting in a so-called Weibull plot. If the BD of N devices is measured, the obtained t_{BD} values are sorted in ascending order, and the ith pair of values of the Weibull plot is given by $t_{BD,i}$ and $F_i = i/(N+1)$. If the data points obey (6.2), they will follow a straight line in the Weibull plot.

A practical example is given in Fig. 6.2 showing the Weibull plot for two unimplanted MOSLEDs without (device 1) and with a SiON layer (device 2). If the BD is intrinsic in nature, i.e., if the defects inducing the BD are created during the electrical operation, the corresponding data points should behave like those of device 1 (open squares). However, defects and other imperfections created during the fabrication process of the device can lead to earlier failures, called extrinsic BD events, which have often a very broad distribution and add to the Weibull distribution [349] as seen in Fig. 6.2 for device 2. In turn, the extrinsic BD events split into those who fail from the very beginning [all data points collected at $\ln(t) = 0$] and those who work in principle but with a reduced lifetime. If the extrinsic data points are subtracted, the remaining data points now obey the Weibull distribution (closed circles). In the present example, the modal lifetimes amount to 64 and 10,800 s

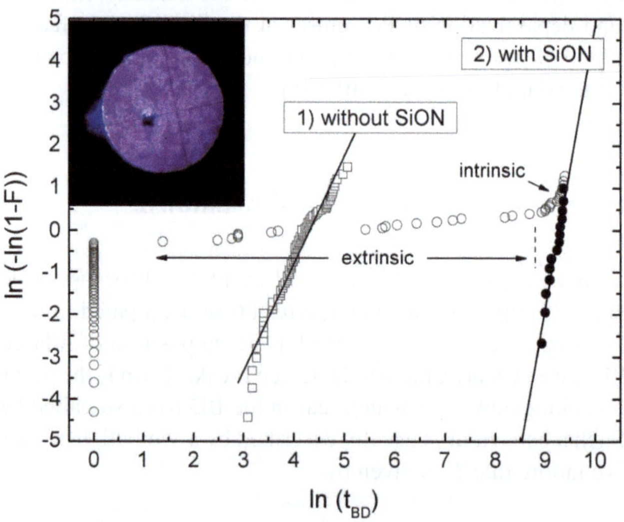

Fig. 6.2 Weibull statistics of unimplanted MOSLEDs with (*open and closed circles*) and without a SiON layer (*open squares*). The *inset* displays an Eu-implanted MOSLED taken at the moment of the BD. Further details are given in the text (after [180])

($Q_{BD} = 0.3$ and $53\,\mathrm{C\,cm^{-2}}$) for device 1 and 2, respectively. More details about BD statistics can be found in [196, 349].

Although the BD of a MOSLED is triggered by a very local event, in case of an ITO electrode it quickly spreads out over the whole electrode area. The snapshot in Fig. 6.2 displays a photograph of an Eu-implanted MOSLED taken at the moment of the BD showing that the whole device is affected. The visual inspection of various MOSLEDs after the BD reveals that the ITO electrode can be both partially and entirely destroyed, but that the latter case occurs more often.

6.1.3 Charge-to-Breakdown Values Under Constant Current Injection

Charge-to-breakdown values were measured under constant current injection and with different injection current densities. Figure 6.3 displays measured t_{BD} values vs. the injection current density for Tb-implanted MOSLEDs processed with LOCOS and with a diameter of 300 μm, a 75 nm thick SiO_2 layer and a 100 nm thick SiON layer. As shown, the measured values obey a reciprocal dependence on the injection current density implying a constant Q_{BD} value of $28\,\mathrm{C\,cm^{-2}}$. As different applied electric fields are necessary to vary the *injection current density* over several orders of magnitude, the degradation of the SiO_2 layer is rather due to the number of injected electrons than caused by the electric field in the range of observed injection current densities.

In consideration of potential applications, the question how the stability of RE-implanted MOSLEDs varies with the fabrication conditions is of special interest. Whereas the beneficial role of an additional SiON layer and the LOCOS processing is discussed in Sect. 6.3, the following discussion focuses on the impact of the annealing temperature, the RE concentration and the effect of different RE elements.

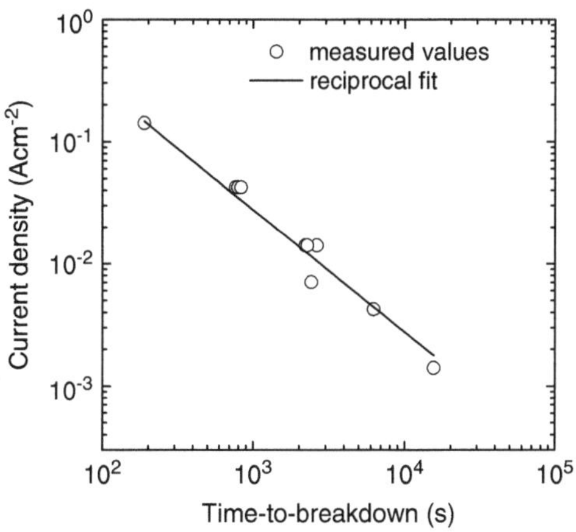

Fig. 6.3 Measured time-to-breakdown values vs. the injection current density for a Tb-implanted MOSLED. The reciprocal fit implies a constant Q_{BD} value of $28\,\mathrm{C\,cm^{-2}}$

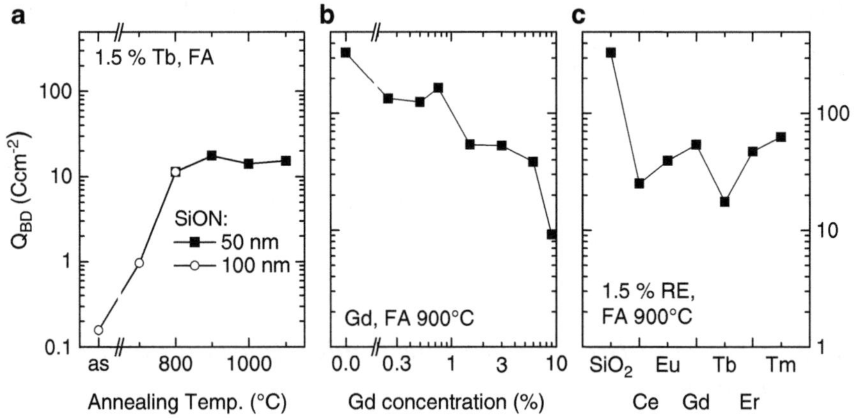

Fig. 6.4 Charge-to-breakdown values of various RE-implanted MOSLEDs determined from a Weibull plot as a function of the annealing temperature (**a**), the Gd concentration (**b**) and the RE element (**c**). The uncertainty is a factor of 2–3. Graph (**a**) has a different scale than (**b**) and (**c**)

Figure 6.4 displays Q_{BD} values of various RE-implanted MOSLEDs with a 100 nm thick SiO_2 layer determined from a Weibull plot as a function of the annealing temperature (a), the Gd concentration (b) and the RE element (c). The Q_{BD} values were determined from a Weibull plot which was based on relatively low sample sizes, for which reason the given Q_{BD} values have an uncertainty of a factor of 2–3.

Undoubtedly, devices annealed at higher annealing temperatures exhibit a higher stability than those which experienced either no annealing or only a low temperature treatment (Fig. 6.4a). This can be easily understood by assuming that the defect density within the dielectrics is the highest for the unannealed case and is supported by the observations that unannealed devices have the highest hydrogen concentration in SiON (Sect. 2.2.2) and show the largest concentration of trapped negative charges in the as-prepared state (Sect. 3.2.1). In addition, as-implanted devices contain a large number of implantation-induced defects which are not yet annealed out. A similar argumentation applies to the RE concentration dependence. As shown in Fig. 6.4b for the case of Gd, the stability of the MOSLEDs continuously decreases with increasing Gd concentration. Compared to the unimplanted case, the RE implantation causes defects (Sect. 2.2.1) which cannot be annealed out completely by FA 900°C. In addition, there are structural defects caused by the presence of Gd in the amorphous SiO_2 network. Clearly, the concentration of these defects will increase with increasing implantation dose and thus increasing Gd concentration. Figure 6.4c compares the Q_{BD} values of MOSLEDs implanted with different RE elements and demonstrates that the implantation of REs with a peak concentration of 1.5% leads to a stability decrease by roughly one order of magnitude. However, the uncertainties of the different Q_{BD} values are too high to decide which RE element gives the best results.

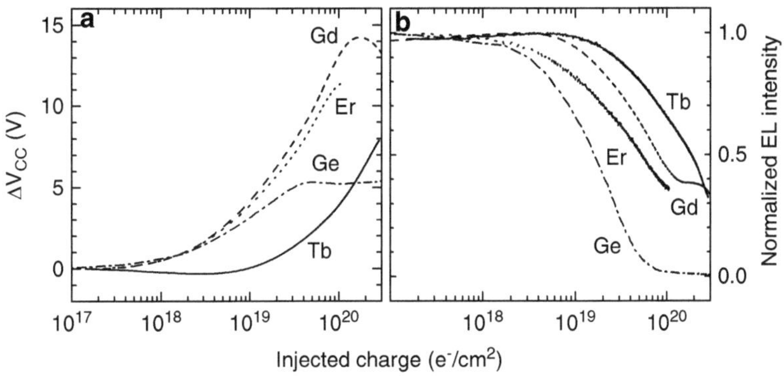

Fig. 6.5 ΔV_{CC} characteristics (**a**) and normalized EL intensity (**b**) of Ge- and RE-implanted SiO_2 layers under constant current injection as a function of the injected charge. The ELI characteristics monitor the main EL line of the corresponding element (after [134])

6.2 Electroluminescence Quenching

6.2.1 The Electroluminescence Quenching Cross-Section

The EL quenching of ion-implanted MOSLEDs was investigated for Ge and the RE elements Gd, Tb and Er in [101] and [134], respectively. Although Ge is not a RE element, it is an ideal candidate for comparison as it shows both similarities and differences compared to the RE-implanted MOSLEDs. As shown in Fig. 6.5, the EL quenching under constant current injection is strongly correlated with charge trapping. The corresponding ΔV_{CC} characteristics measured at an injection current density of $0.1\,\mathrm{A\,cm^{-1}}$ (Fig. 6.5a) are characterized by an onset of strong electron trapping for an injected charge of 10^{18}–$10^{19}\,\mathrm{e^-\,cm^{-2}}$ (phase III of charge injection, Sect. 3.4.3), which changes into hole trapping just before breakdown for an injected charge higher than $10^{20}\,\mathrm{e^-\,cm^{-2}}$. Concurrently, the ELI characteristics – the normalized EL intensity as a function of the injected charge – are on a constant level up to 10^{18}–$10^{19}\,\mathrm{e^-\,cm^{-2}}$, followed by a drop down for higher injection currents (Fig. 6.5b).

As outlined in [101], the EL quenching can be described with a similar first-order rate equation as the charge trapping described in Sect. 3.1.3. According to this, the number of potential luminescence centres N_{LC} decreases with a constant decay rate $k = \tau_Q^{-1}$ with τ_Q being the characteristic EL quenching time. This leads to the rate equation

$$\dot{N}_{LC} = -k\, N_{LC} \tag{6.4}$$

and gives, if the times are transformed into injected charges, the simple solution

$$N_{LC} = N_{LC}^0 \exp\left(-\frac{\sigma_Q\, Q_{inj}}{q}\right), \tag{6.5}$$

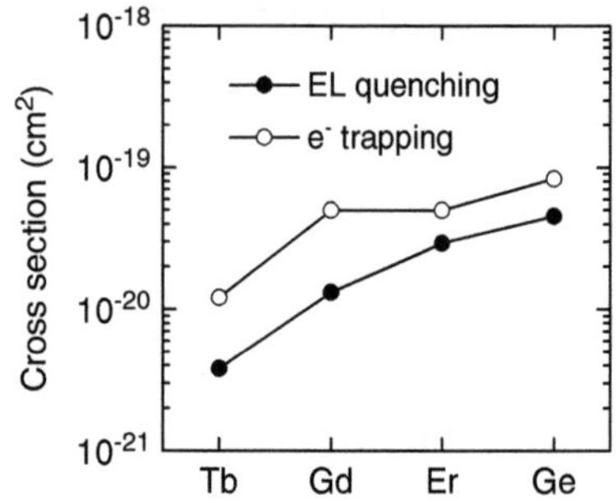

Fig. 6.6 EL quenching and electron trapping cross-section of Ge- and RE-implanted MOSLEDs determined from the characteristics shown in Fig. 6.5

with N^0_{LC}, σ_Q and q being the initial number of LCs, the quenching cross-section and the elementary charge, respectively. Using (6.5) and (3.4), both the EL quenching and electron trapping cross-section can be calculated from the experimental characteristics shown in Fig. 6.5 resulting in the dependency displayed in Fig. 6.6. Both types of cross-sections increase in the order Tb, Gd, Er and Ge strongly implying that the EL quenching is caused by electron trapping.

Close inspection of Fig. 6.5b reveals that the EL intensity of the Gd-related characteristic decreases by 60% before it approaches a short interval of constant EL intensity. This plateau is reached at $Q_{inj} > 10^{20}\,e^-\,cm^{-2}$ where the corresponding ΔV_{CC} characteristic is at maximum marking the transition from electron trapping to positive charge accumulation (Fig. 6.5a). This is a first indication that the EL quenching process is a two-stage process. Figure 6.7 displays the ΔV_{CC} and ELI characteristic monitored at 541 nm of a long-living Tb-implanted MOSLED where the two-stage EL quenching process is more obvious. Similar to the case of Gd in Fig. 6.5, the EL intensity drops down by 60% and reaches a temporary, albeit very small plateau, before the final or second stage of quenching starts.

6.2.2 The Electroluminescence Quenching Model

Based on the previous chapter, the EL quenching model for RE-implanted MOSLEDs can now be developed. In general, EL quenching can be ascribed to one of the following cases:

1. The number of *excited* RE^{3+} ions remains unchanged, but an additional non-radiative decay path quenches the EL.
2. The number of *excited* RE^{3+} ions decreases. This is equal to a decrease of the excitation cross-section with ongoing charge injection.
3. The *total* number of RE^{3+} ions decreases.

The first possibility was investigated by measuring the decay time of the green EL line of Tb-implanted MOSLEDs from the virgin, unstressed device and after injecting an increasing number of charges. Although the EL intensity decreases, the EL decay time shows only insignificant changes with Q_{inj}, excluding the possibility of an additional non-radiative decay path [144]. The other two possibilities are more difficult to distinguish. Whereas the decrease of the number of excited RE^{3+} ions points to some deactivation or blocking of them, a decrease of the total number of RE^{3+} ions implies the transformation of them during constant current injection. However, as it will be shown in Sect. 6.2.3, the EL quenching of RE-implanted MOSLEDs results mainly from the increasing blocking of RE^{3+} ions with increasing charge injection.

According to the defect shell model (Sect. 3.4.1), the REO_x clusters are surrounded by a shell of traps and defect precursors. The high-level charge injection is characterized by a strong electron trapping in the bulk of the SiO_2 layer accompanied by a strong accumulation of positive charges at the Si–SiO_2 interface (Sect. 3.4.3). Due to the continuous impact of hot electrons, weak and strained Si bonds in the defect shells can break and serve as electron traps. Other defect precursors can also be transformed into defects and electron traps as well. Thus, the REO_x clusters and the RE^{3+} ions are increasingly surrounded by trapped electrons which screen the RE^{3+} ions from hot electrons by Coulomb repulsion. In addition, the continuous defect creation leads to an additional scattering of the hot electrons in the CB of SiO_2 and thus to a lowering of their kinetic energy. This in turn lowers the excitation cross-section of the RE^{3+} ions as the fraction of hot electrons having enough kinetic energy for impact excitation decreases. This scenario is strongly supported by the coincidence of electron trapping with the first stage of the EL quenching in Fig. 6.7.

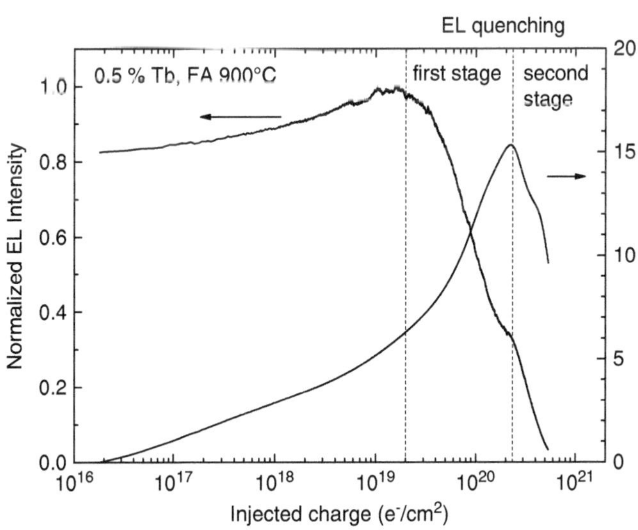

Fig. 6.7 ΔV_{CC} and ELI characteristic of a MOSLED with 0.5% Tb and annealed by FA 900°C. The two stages of EL quenching can be clearly seen. The ELI characteristic was monitored at 541 nm under an injection current density of 4×10^{-3} A cm^{-2} (after [214])

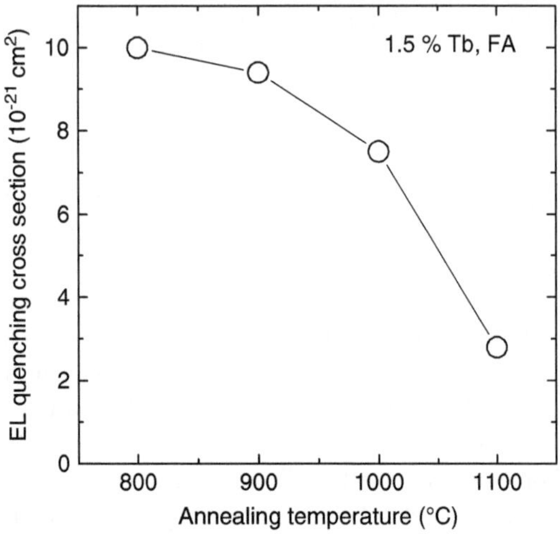

Fig. 6.8 EL quenching cross-section of Tb-implanted MOSLEDs as a function of the annealing temperature (after [214])

The second stage of EL quenching starts when the positive charges at the Si–SiO_2 interface have accumulated to such an extent that their region expands beyond the tunnelling distance from the Si–SiO_2 interface. Positive charges at this location now effectively lower the injection barrier just by its Coulomb attraction which causes the decrease of the ΔV_{CC} characteristic for $Q_{inj} > 2 \times 10^{20}\,e^{-}\,cm^{-2}$ in Fig. 6.7. As the physical processes of defect creation and electron trapping in the second stage of EL quenching are basically the same as in the first one, the EL intensity continues to decrease. Only the transition between the two stages gives a short period of a decelerated EL quenching, most probably caused by a temporary and partial charge recombination of the trapped electrons.

Finally, the EL quenching seems to scale with the size of the REO_x clusters. Figure 6.8 displays the EL quenching cross-section of Tb-implanted MOSLEDs which decreases with increasing annealing temperature. This tendency correlates well with the increase of the average REO_x cluster size (Fig. 2.12) and thus with an increase of the defect shell around the clusters. Clearly, larger REO_x clusters exhibit less EL intensity (Sect. 4.4.1), but this smaller intensity is quenched by a smaller quenching cross-section. It is assumed that the defect shells increase with increasing REO_x cluster size and coalesce with each other forming a broad defect region in the SiO_2 matrix. Although the total number of defects might be higher for larger clusters, the defects are more dispersed in the defect region. This in turn can decrease the action of Coulomb repulsion leading to a later onset of the EL quenching [214].

6.2.3 *The Electroluminescence Reactivation Experiment*

This experiment was designed in order to study the physical nature of the EL quenching mechanism and was performed at Tb-, Tm- and Ge-implanted MOSLEDs

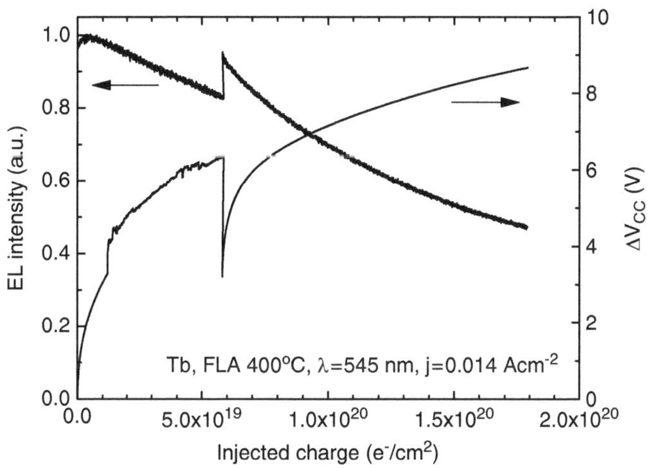

Fig. 6.9 ELI and ΔV_{CC} characteristic as a function of the injected charge for a Tb-implanted MOSLED (after [144])

with a 100 nm thick SiO_2 and a 50 nm thick SiON layer [144]. The general experimental regime can be followed in Fig. 6.9 showing the ELI and ΔV_{CC} characteristic as a function of the injected charge under constant current injection. The constant current injection was stopped when the EL intensity dropped to approximately 80% of its initial value. Subsequently, the device was exposed to FLA 400°C 20 ms (in order to not exceed the temperature of the Al contact anneal), followed by the resumption of the constant current injection. Whereas the ΔV_{CC} value drops down by at least 3 V, the EL intensity nearly recovers to its initial value. Further charge injection leads once again to EL quenching and an increase of ΔV_{CC} indicating a continuing electron trapping.

These observations can be interpreted as follows: During charge injection, electrons are trapped in the defect shell around the TbO_x clusters and in the vicinity of Tb^{3+} ions which results in an efficient screening of this LCs by Coulomb repulsion. Applying FLA leads to a release of most of the trapped electrons; as a consequence, the reactivated Tb^{3+} ions are available for excitation again. However, a close inspection of the ELI and ΔV_{CC} characteristic after FLA reveals that both the EL quenching and the charge trapping are now accelerated which compensates the positive effect of the FLA in the long term. This is due to the fact that the charge injection is accompanied by the creation of new defects which are not removed by the moderate FLA treatment. Therefore, charge injection after the FLA reactivation starts with a higher defect density allowing a higher rate of EL quenching and charging.

Similar results were obtained for Tm-implanted, but not for Ge-implanted SiO_2 layers. Figure 6.10 shows the ELI characteristic for the 390 nm line of a Ge-implanted MOSLED which is most probably caused by a ≡Ge–Ge≡ defect [66, 144]. In contrast to the RE-related LCs, the post-injection FLA leads only to a

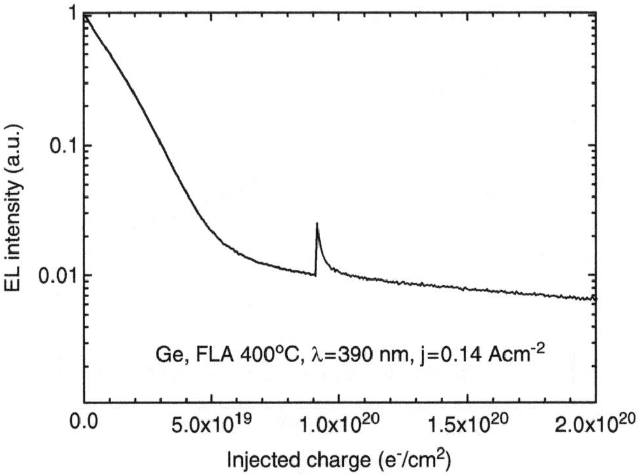

Fig. 6.10 Normalized EL intensity as a function of the injected charge for a Ge-implanted MOSLED

slight and not sustainable recovery of the EL intensity. It is assumed that electrons are directly trapped by NOV defects leading to bond breaking and thus to a transformation of the NOV centres into optically non-active defect centres. Moreover, bond breaking may also occur as a result of impact ionization of hot electrons. Because of these reasons, the EL quenching in case of Ge is mostly irreversible. More details to charge trapping and EL quenching in Ge-implanted MOSLEDs can be found in [101, 212, 350]. It should be noted that similar results were obtained for the weak EL of unimplanted MOSLEDs by monitoring the EL line around 470 nm [144] most probably caused by a NOV (≡Si–Si≡). In this case, the same argumentation applies as for the ≡Ge–Ge≡ defect.

Finally, the question remains whether the successful reactivation in case of RE-implanted MOSLEDs is more due to the photon irradiation or the thermal budget applied during the FLA treatment. Figure 6.11 compares the efficiency of the RE reactivation process for different annealing types and temperatures. After post-injection FA at 400°C for 1 min, only a slight increase of the EL intensity was observed. The highest efficiency of the reactivation process was obtained after FLA at 400°C for 20 ms for both Tb- and Tm-implanted samples. It was found that RE implantation into the oxide layer creates electron traps lying around 3 eV below the CB of the SiO_2 [134]. Therefore, the thermal energy supplied to the trapped electrons during FA at 400°C is not sufficient to release them from the traps. In the case of FLA, however, the devices are exposed to an intense flash whose maximum at the wavelength scale is in the blue-green spectral region. Thus, electron detrapping may occur via photon absorption, which is assisted by the thermal budget of the FLA treatment. In summary, the EL reactivation experiment strongly supports the idea that the blocking of RE^{3+} ions by the Coulomb repulsion of nearby trapped electrons is the main EL quenching mechanism.

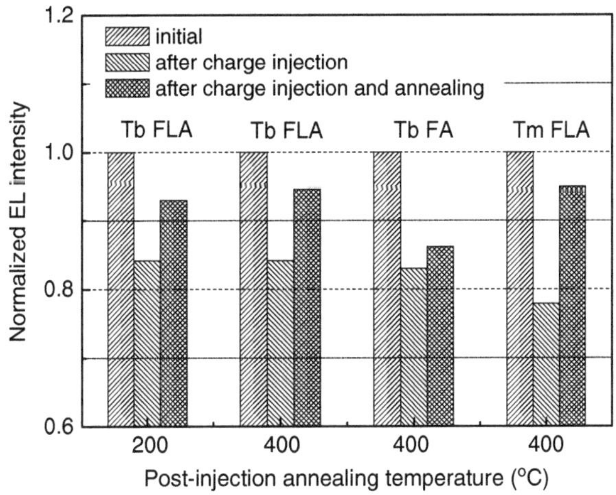

Fig. 6.11 Initial EL intensity (*first bar*), EL intensity after charge injection (*second bar*) and after charge injection and annealing (*third bar*). The EL was measured under the same constant current injection at 10 μA for Tb and 50 μA for Tm (adopted from [144])

6.2.4 Temperature-Dependent Electroluminescence Quenching

As already mentioned, at elevated temperatures degradation processes take place in an accelerated way, which are most probably the result of defect activation [342] and enhanced defect generation. Therefore, the investigation of the temperature-dependent EL quenching may reveal further details about its nature and correlation with charge trapping. Figure 6.12a displays the ΔV_{CC} characteristics of a Gd-implanted MOSLED under constant current injection for different operation temperatures measured either on a heated chuck (300–423 K) or in a cryogenic chamber (20–220 K). Obviously, the ΔV_{CC} characteristics shift to higher injected charges with lower operation temperatures implying a lower electron trapping. However, the characteristics measured in the cryogenic chamber are disturbed by electronic noise resulting in an additional voltage jitter, for which reason the presented curves for 220 and 70 K in Fig. 6.12a are smoothed. Concurrently, the EL intensity of the 316 nm line was monitored resulting in the EL quenching curves shown in Fig. 6.12b. Similar to charge trapping, EL quenching is retarded with decreasing operation temperature.

In general, electron trapping and EL quenching cross-sections can be estimated from the data shown in Fig. 6.12. However, the measured characteristics are not ideal which introduces a slight degree of arbitrariness in the form of additional assumptions necessary in order to model these non-ideal curves in a satisfying way. First of all, there is not a single cross-section but a distribution of cross-sections, for which reason estimated cross-section values are only average values. In addition, (3.3) and (3.4) assume a constant trap density which, however, increases with increasing charge injection due to defect creation. This is most apparent for

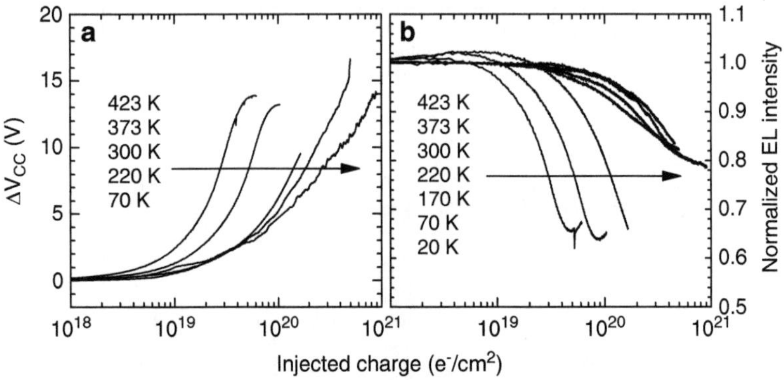

Fig. 6.12 ΔV_{CC} characteristics (**a**) and ELI characteristics (**b**) of a Gd-implanted MOSLED under constant current injection for different operation temperatures (original data from [351]). The *arrows* point from higher to lower operation temperatures. More details are given in the text

373 and 423 K where two different trapping cross-sections are required for modelling the corresponding ΔV_{CC} characteristics. Whereas the lower cross-section (type I in Fig. 6.13) dominates for lower injected charges, charge trapping is accelerated for $Q_{inj} > 6\times10^{19}\,e^{-}\,cm^{-2}$ and $Q_{inj} > 3\times10^{19}\,e^{-}\,cm^{-2}$ for 373 and 423 K, respectively. This acceleration is modelled by a correspondingly larger trapping cross-section (type II in Fig. 6.13). In case of EL quenching, a constant offset was assumed in order to model the transition between the two stages of EL quenching (Sect. 6.2.2). As a consequence, the cross-section values obtained from the fitting procedures are assumed to have an uncertainty in the order of 50%. These cross-sections are displayed in Fig. 6.13 showing the tendency to increase with increasing operation temperature. Especially in the range of 220 and 473 K, both the charge trapping and EL quenching cross-section run parallel underlying the close correlation between the two phenomena. However, the uncertainties are too large to allow the serious estimation of activation energies from these temperature dependencies.

Finally, a few words on the physical background of the temperature dependence of the EL quenching: Measuring the EL decay time as a function of temperature reveals only small variations around an average value of 1.43 ms for 2% Gd and FA 900°C [351]. Thus, the enhanced electron trapping with increasing operation temperature does not primarily introduce an additional non-radiative decay path for the relevant LCs, but rather quenches the EL by Coulomb blocking of the active LCs.

6.2.5 The Anomalous Electroluminescence Quenching Behaviour of Eu

Under certain conditions, the EL quenching of Eu-implanted MOSLEDs exhibits an unusual dependence on the injected charge, namely, an initial strong increase

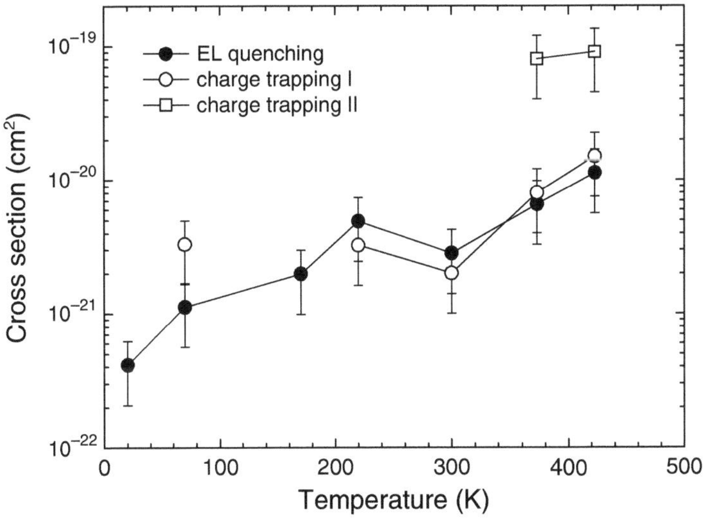

Fig. 6.13 EL quenching (*solid symbols*) and charge trapping cross-sections (*open symbols*) as derived from the data in Fig. 6.12. More details are given in the text

of the EL intensity followed by the known quenching behaviour described in the previous chapters. The evolution of the EL spectrum with the injected charge was investigated for MOSLEDs with Eu peak concentrations of 0.1, 0.5 and 1.5% which were annealed with FA at different temperatures [132]. As shown in Fig. 6.14a, the shape of the EL spectrum does not change very much with increasing charge injection in case of FA 800°C. However, the intensity of the red EL caused by Eu^{3+} at 618 nm increases up to an injected charge of $2 \times 10^{19}\,e^- cm^{-2}$, followed by a smooth decrease for higher injected charges. In maximum, the gain of EL intensity compared to the initial value amounts to approximately 30%. Sporadically, such a behaviour has also been observed for other RE elements, e.g., for Tb in Fig. 6.7. A similar behaviour can be observed for FA 850°C and 900°C and for concentrations of 0.5% and 0.1% (not shown), but at 950°C the figure will change. The unstressed spectrum shows a broad EL band of low intensity extending from the blue-green up to red with a small Eu^{3+} peak on top of it (Fig. 6.14b, unstressed spectrum). With continuing charge injection, the broad EL band gradually vanishes, and the red Eu^{3+} peak increases with time. This development is accompanied by the appearance of a blue-violet peak around 415 nm possibly caused by Eu^{2+}. In case of FA 1,000°C (Fig. 6.14c), the red Eu^{3+} peak is hard to detect at the beginning, but starts to rise with time, too. Unfortunately, the stability of the device was not high enough to observe a more spectacular outgrow of the red EL. It has to be noted that a similar but weaker behaviour as in Fig. 6.14b can be observed already in the case of FA 900°C, 0.1%. It is therefore assumed that the broad EL band is generally present at FA 900°C, but with minor intensity and superimposed by the much stronger signal from Eu^{3+} in case of 0.5 and 1.5% Eu. As already discussed in Sect. 4.1.4.3,

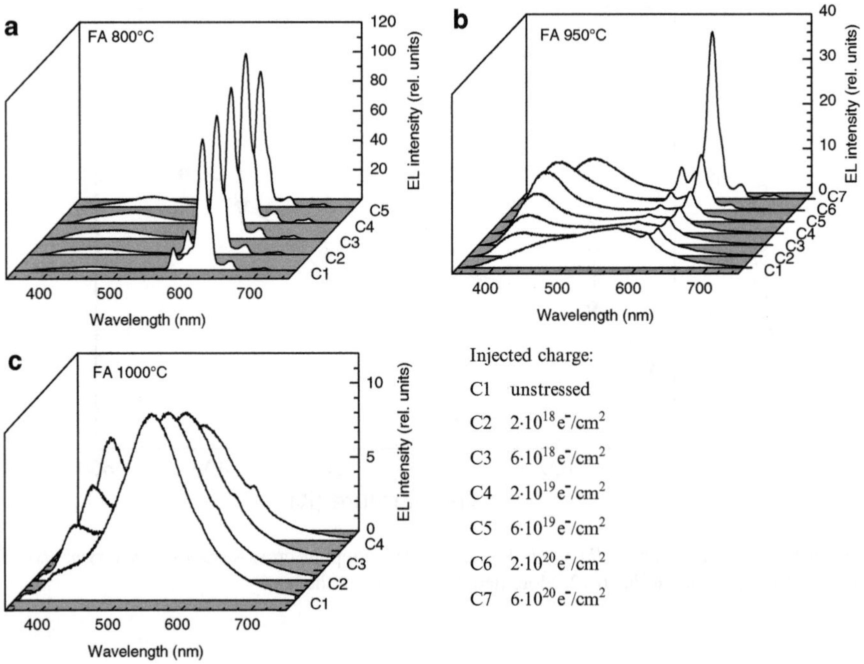

Fig. 6.14 EL spectrum of Eu-implanted MOSLEDs containing 1.5% Eu and annealed with FA at 800°C (**a**), 950°C (**b**) and 1,000°C (**c**) for various levels of charge injection. The EL spectra were taken during charge injection under a constant current density of 7×10^{-3} A cm^{-2} (after [132])

the broad EL band is possibly composed of contributions from various defects in ion-implanted SiO_2 and from Eu^{2+}.

Figure 6.15 shows the integral over the red Eu^{3+} peak at 618 nm as a function of the injected charge for an injection current density of 7×10^{-3} A cm^{-2} (5 μA) for different annealing temperatures (a) and different Eu concentrations (b). The integral was corrected by background subtraction which is of special importance in case of higher annealing temperatures where the Eu^{3+} signal is small. In general, the integrated EL intensity of the red Eu^{3+} peak, also called the lifetime characteristics, increases from an initial value EL_0 with increasing injected charge, reaches a maximum at an injected charge Q_{inj}^{max} and is finally quenched for very high injected charges. If the ratio between the maximum EL intensity achieved at Q_{inj}^{max} and EL_0 is denoted as θ, the following tendencies can be observed: With increasing anneal temperature, EL_0 decreases strongly but is accompanied by a strong increase of θ which can reach a value of 100 in case of 0.1% Eu and FA 950°C as shown in Fig. 6.15b. With increasing Eu concentration, EL_0 increases with an adequate decrease of θ. Whether there is a dependence of Q_{inj}^{max} on the Eu concentration is difficult to say, as in many cases, the devices were not stable enough to show the whole range of the lifetime characteristic.

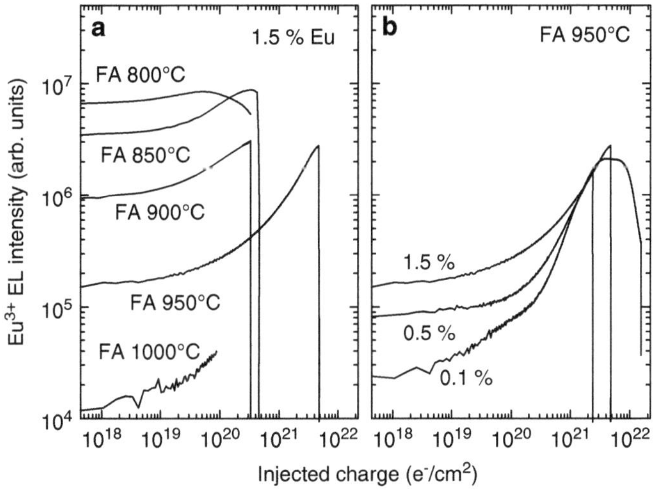

Fig. 6.15 Integrated Eu^{3+} EL intensity as a function of the injected charge for devices implanted with 1.5% Eu and annealed at different temperatures (**a**) and for devices annealed with FA 950°C having different Eu concentrations (**b**) (after [132])

The next question is to which extent the lifetime characteristic of the Eu-implanted MOSLEDs depends on the injection current. Figure 6.16a displays the lifetime characteristics of such a device with 0.1% Eu for different injection current densities and normalized with respect to the maximum EL intensity for better projection. First of all, the general shape of the lifetime characteristic is not changed very much, but there is a shift of $Q_{\text{inj}}^{\text{max}}$ to smaller injected charges with increasing injection currents. However, this shift is quite small: the increase of the current density by a factor of 20 leads to a shift of $Q_{\text{inj}}^{\text{max}}$ by a factor of slightly more than 2. Apparently, with increasing injection current, an enhanced light emission is observed, but the increase of EL_0 is accompanied by a decrease of θ. The corresponding ΔV_{CC} characteristics shown in Fig. 6.16b illustrate that the rise phenomenon is accompanied by strong electron trapping and occurs in an injected charge region ($>10^{19}\,\text{e}^-\,\text{cm}^{-2}$) where the normal EL quenching starts.

Another interesting feature can be observed by comparing the lifetime characteristics of devices processed with different types of annealing as shown in Fig. 6.17. For FLA 900°C, the curve starts with a high EL_0 value but decreases continuously with increasing injected charge. For RTA 900°C, 60 s a maximum at $Q_{\text{inj}}^{\text{max}} \approx 10^{20}\,\text{e}^-\,\text{cm}^{-2}$ can be observed which is shifted to larger injected charges for FA 900°C. It seems that the anomalous behaviour of the lifetime characteristic is closely linked with high thermal budgets, whereas low thermal budgets like FLA lead to the usual quenching behaviour of RE-implanted MOSLEDs.

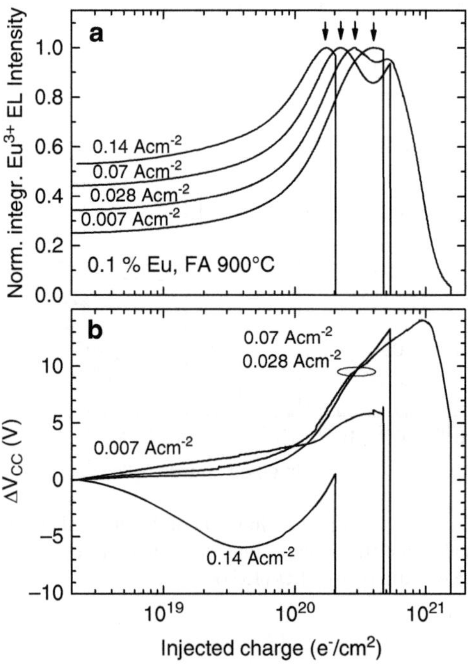

Fig. 6.16 Normalized, integrated Eu^{3+} EL intensity (**a**) and the change of the applied voltage (**b**) as a function of the injected charge for FA 900°C and 0.1% Eu and its dependence on the injection current density (after [132]). The maxima of the EL intensity are marked with *small arrows*. One of the ΔV_{CC} characteristics was corrected for voltage jitter

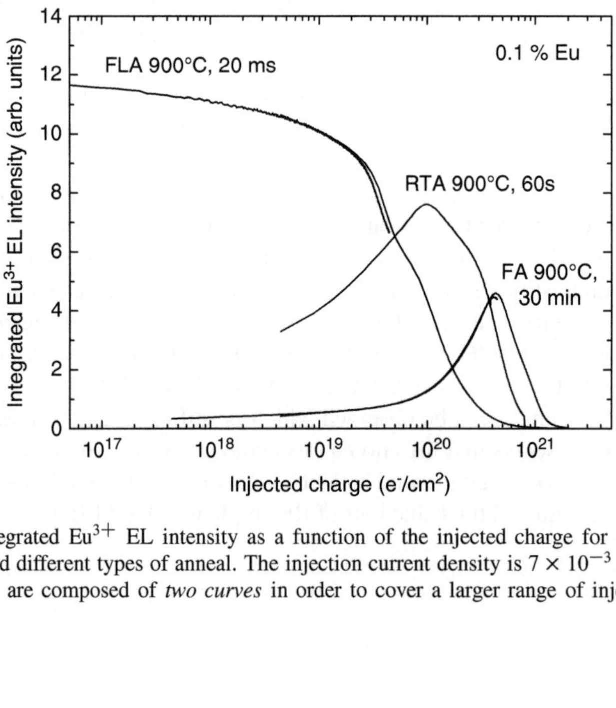

Fig. 6.17 Integrated Eu^{3+} EL intensity as a function of the injected charge for Eu-implanted MOSLEDs and different types of anneal. The injection current density is 7×10^{-3} A cm^{-2}. Two characteristics are composed of *two curves* in order to cover a larger range of injected charges (after [132])

6.2.6 Qualitative Model of the Electroluminescence Rise Phenomenon

If a rise of EL intensity with increasing injected charge is observed, the possible mechanism can be assigned in principle to one of the following categories analogous to Sect. 6.2.2:

(a) The number of Eu^{3+} ions is increasing during injection.
(b) A non-radiative decay path changes in intensity and decreases the degree of EL quenching during injection.
(c) The excitation of the Eu^{3+} ions becomes more efficient during injection.

Starting the discussion with category (a), it has to be stated first that all structural processes requiring high temperatures, namely RE diffusion and REO_x cluster formation, are unlikely to occur under high field charge injection. It has to be assumed that the microstructure of the unstressed device in terms of Eu distribution, cluster size and interface agglomerations remains unchanged during the whole period of charge injection. However, the number of Eu^{3+} ions may change due to recharging processes. Indeed, Eu can be present both in the divalent and trivalent ionic state (Sect. 4.1.4.3), and depending on the Eu concentration and the annealing conditions a partially strong blue EL was found in Eu-implanted structures which was attributed to Eu^{2+} ions [132]. If this scenario would be of significance, Eu^{2+} ions have to transform into Eu^{3+} ions in order to cause the observed rise of the red EL intensity. There are two counter-arguments. First, the unstressed structure in our case should exhibit a strong blue EL, especially in the case of FA 950°C, which was not observed. Second, during the rise phase, a concurrent increase of *both* the blue and red EL was observed for some devices.

EL decay time measurements are the most suitable method to check if a change in the luminescence intensity is caused by the change of non-radiative decay paths (category b). Figure 6.18 shows that the decay times of the unstressed devices exhibit a weak but significant dependence on the annealing temperature. In Fig. 6.19, the EL decay time values after certain levels of charge injection are given and correlated with the corresponding ELI characteristic. If the EL is quenched by non-radiative decay paths, the corresponding decay time should decrease by the same factor as the EL intensity does. However, in our case, decay times between 580 and 730 μs for the red $^5D_0$–$^7F_2$ line are measured, although the EL intensity drops down by a factor of nearly 40 if the injected charge increases from 2×10^{20} to 8×10^{20} $e^- cm^{-2}$. Moreover, close inspection of Fig. 6.19 reveals that the EL intensity increases up to an injected charge of 2×10^{20} $e^- cm^{-2}$, whereas the decay time shortens from 711 to 588 μs. This contradicts the idea that an increase of non-radiative decay paths simultaneously causes a shortening of the EL decay time and a decrease of EL intensity (Sect. 5.1.3). It is therefore concluded that non-radiative decay paths give only a minor contribution to the observed rise phase of the EL spectrum evolution.

The remaining potential mechanism (category c) for the EL rise phenomenon, namely, the increase of the Eu^{3+} excitation efficiency during the rise phase, should

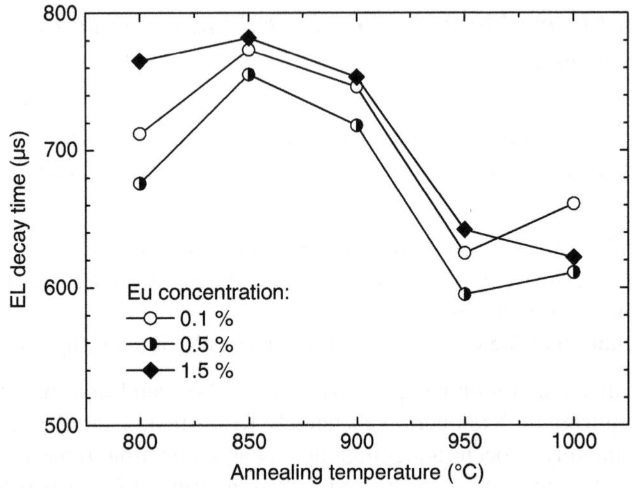

Fig. 6.18 EL decay times of Eu-implanted MOSLEDs as a function of Eu concentration and annealing temperature (after [132])

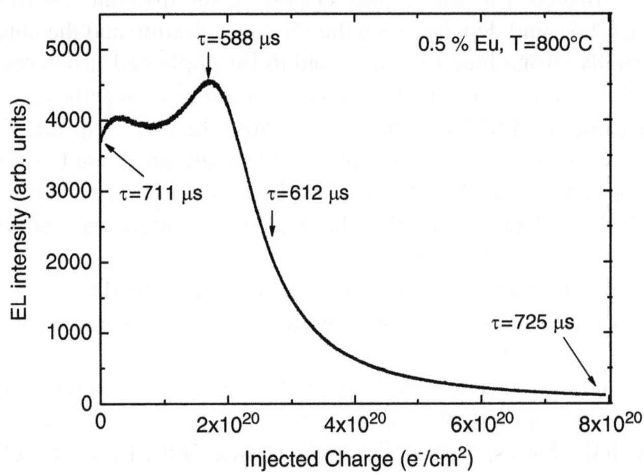

Fig. 6.19 EL intensity of an Eu-implanted MOSLED annealed by FA 800 ° C and containing 0.5% Eu. After certain levels of charge injection EL decay time measurements were performed whose decay times are assigned by *arrows* to the corresponding points of the ELI characteristic (after [132])

now be elucidated in detail. The present data are not sufficient to verify a definite mechanism, but based on the correlations between microstructure and electrical and optical properties, a hypothesis about the most probable mechanism can be developed. A close look to Fig. 2.11 reveals that at FA 900°C and RTA 1,000°C 6 s, a considerable diffusion of Eu towards the interfaces of the gate oxide layer starts which culminates in strong interface decorations and large clusters up to 20 nm in case of FA 1,000°C (Figs. 2.13 and 2.14). Concurrently, the rise phase of the EL intensity is more pronounced for high thermal budgets (Figs. 6.15 and 6.17). This leads at first to the question at which location excitable Er^{3+} ions can be found in the oxide layer.

In [190], it was reported that luminescent Eu^{3+} ions are either finely dispersed in the oxide matrix or located close to or at the EuO_x cluster surface. With increasing thermal budget, the number of luminescent Eu^{3+} ions decreases as the dispersed Eu^{3+} ions will diffuse towards the interfaces or the growing clusters. At the same time, the surface-to-volume ratio of the clusters decreases, and in turn a general decrease of EL intensity is observed with increasing annealing temperature. For 950°C and 1,000°C, it is assumed that nearly all Eu^{3+} ions are located at the oxide interfaces or part of the clusters, and that during the rise phase the excitation conditions are improved for them.

The situation for the Eu-implanted MOSLED annealed with a high thermal budget at the beginning of charge injection (a) and in an advanced state of charge injection (b) is illustrated in Fig. 6.20. At the bottom, the RBS profile for FA 1,000°C (see also Fig. 2.11) is given (c) showing three different regions 1, 2 and 3 with Eu segregations in the interface regions 1 and 3 and large EuO_x clusters in region 2. The capability of the broad defect region developed from the coalesced defect shells of the large clusters (Sect. 6.2.2) is indicated by an extended distribution of traps in the oxide layer. At the beginning of charge injection, these traps

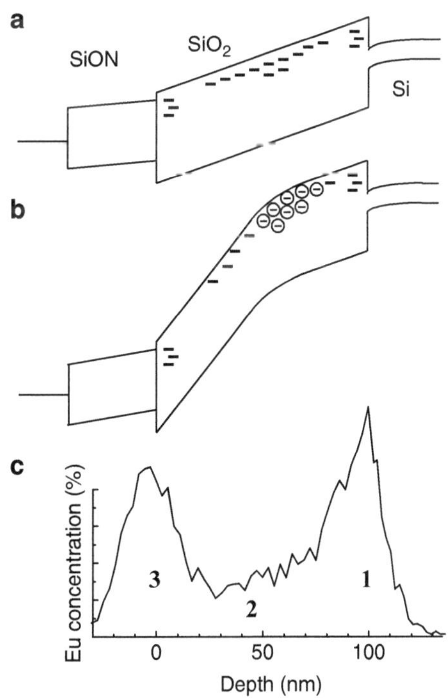

Fig. 6.20 Schematic band diagram of an Eu-implanted MOSLED under forward bias at the beginning of (**a**) and after considerable charge injection (**b**). The Eu profile for FA 1,000°C is given at the *bottom* (**c**) (after [132])

are assumed to be empty (Fig. 6.20a) or even positively charged. With increasing charge injection, more and more electrons are trapped by the clusters which leads to a down bending of the band diagram between region 2 and 3 as the electric field between region 1 and 2 is kept constant in order to ensure a constant injection current (Fig. 6.20b). Thus, the electric field between region 2 and 3 is enhanced.

The effect on potential LCs being located in the different Eu-rich regions is the following. It is assumed that Eu^{3+} ions in region 1 cannot be excited by the injected electrons as they are located within the dark zone (Sect. 5.2.2) and do not have enough kinetic energy. Considering the trapping of negative charges, Eu^{3+} ions residing at the cluster surface or in the core of *negatively* charged clusters are also unsuitable candidates for causing the EL rise phase, as they are effectively screened from hot electrons by Coulomb repulsion, albeit this screening is damped in case of large REO_x clusters (Sect. 6.2.2). However, this mostly applies to clusters more close to the Si–SiO_2 interface. As the electric field increases towards the SiO_2–SiON interface due to band bending, a part of the clusters of region 2 is also located in the high field region. Eu^{3+} ions in region 3 and partly from region 2 profit from the enhanced local electric field in two aspects. First, the enhanced electric field causes a shift of the energy distribution of hot electrons to higher energies, which in turn increases the probability for an electron to excite an Eu^{3+} ion by impact excitation. Second, enhanced detrapping and field ionization may occur. In this scenario, most of the Eu^{3+} ions at the SiO_2–SiON interface or in the clusters are blocked by trapped electrons in the close vicinity. Higher electric fields shift the balance between charge trapping and detrapping to a lower occupancy of the traps [222]. As a consequence, less and less Eu^{3+} ions are blocked by electrons which results in an increase of EL intensity. As in some cases, an increase of both the blue and red EL lines is observed [352], the direct transformation $Eu^{2+} \rightarrow Eu^{3+}$ triggered by the high electric field is unlikely.

The EL rise phenomenon occurs only under high-level charge injection which is associated with the normal EL quenching. However, the EL rise phenomenon can only delay but not prevent the EL quenching as illustrated in Fig. 6.16: The lifetime characteristics recorded under $0.028\,\mathrm{A\,cm^{-2}}$ finally declines for $Q_{\mathrm{inj}} > 6 \times 10^{20}\,\mathrm{e^-\,cm^{-2}}$ which is well before $Q_{\mathrm{inj}} = 10^{21}\,\mathrm{e^-\,cm^{-2}}$ where the ΔV_{CC} characteristic is at maximum, indicating the transition between the first and second stage of EL quenching in normal cases. Figure 6.21 displays the ELI characteristics (a) and the normalized applied voltage (b) of Eu-implanted MOSLEDs annealed by FA 800°C and FA 950°C. The characteristics were measured under an injection current density of $5 \times 10^{-3}\,\mathrm{A\,cm^{-2}}$ and at an elevated temperature of 110°C where the EL quenching processes are accelerated (Sect. 6.2.4). Under such conditions, a good correlation between the onset of EL quenching (maximum of the ELI characteristic) and the onset of positive charge accumulation near the Si–SiO_2 interface (maximum of the normalized V_{CC} characteristic) can be observed. Thus, in case of large EuO_x clusters, the first stage of EL quenching is reduced or suppressed due to the EL rise phenomenon. However, the second stage of EL quenching occurs similar to other RE-implanted MOSLEDs, albeit at higher injected charges.

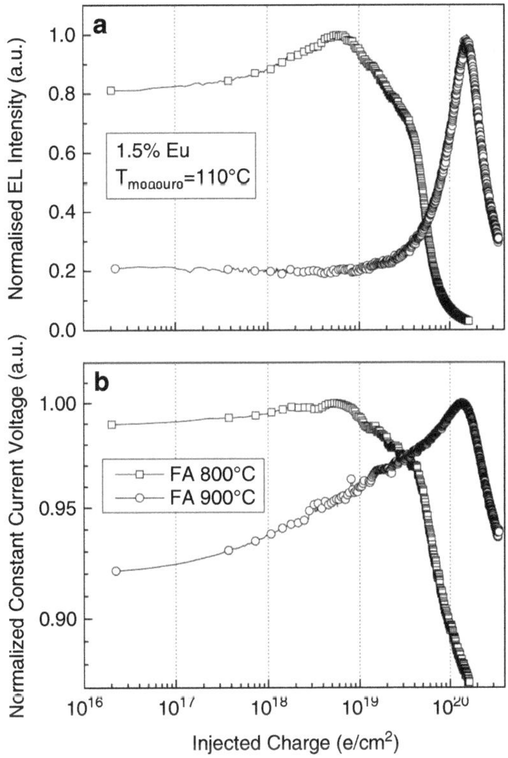

Fig. 6.21 ELI characteristics (**a**) and the normalized applied voltage (**b**) of Eu-implanted MOSLEDs measured at an elevated temperature of 110°C. There is a good correlation between the onset of EL quenching and the onset of positive charge accumulation near the Si–SiO_2 interface (after [214])

It is conceivable to apply such a scenario also to other RE elements if the microstructural conditions comply with that of Eu, namely with the existence of large RE or RE oxide clusters and a massive redistribution of the implanted RE ions towards the oxide interfaces. However, for other RE elements, the EL rise phenomenon can be observed only in a rudimentary form, even for Tb annealed by FA 1,100°C. Therefore, the question whether Eu gives an additional, intrinsic contribution to the observed phenomenon cannot be finally answered.

6.3 LOCOS Processing and the Use of Dielectric Buffer Layers

LOCOS processing is a widely used technique in semiconductor industry to avoid an early electrical BD of SiO_2 layers. As known the electric field at sharp edges is enhanced resulting in potentially weak spots at these locations. This is illustrated by the fact that devices with and without LOCOS processing after breakdown preferably show a decomposition of the ITO electrode on the whole area and at a local spot on the edge of the electrode, respectively. In addition, higher Q_{BD} values are measured in case of LOCOS processing.

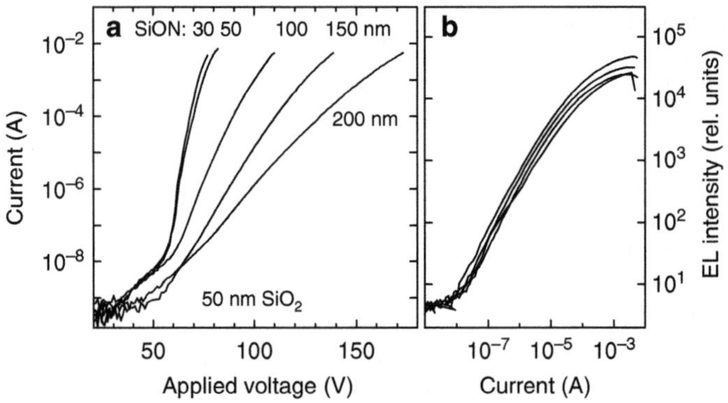

Fig. 6.22 *IV* characteristics (**a**) and the EL intensity dependence on the injection current (**b**) for Tb-implanted devices with a 50 nm thick gate oxide and a SiON layer of different thickness. The Tb peak concentration amounts to 2%, and the devices were annealed with FA 900°C (after [142])

In order to investigate the beneficial role of a SiON protection layer, Tb-implanted MOSLEDs with a 50 nm thick gate oxide and a SiON layer of different thickness were fabricated and characterized. The corresponding *IV* characteristics and EL dependency curves (Fig. 6.22) reveal that the onset of FN tunnelling and the onset of EL emission are nearly the same for all devices, but that the *IV* characteristics become shallower with increasing SiON thickness. Concurrently, the EL intensities of the different devices are close together allowing the tailoring of the SiON thickness to the needs of a potential application without a substantial loss of EL intensity. The small differences are mainly caused by the additional voltage drop across a thick SiON layer in order to have the same electric field at the injecting interface.

In [142], it was found that the best stability is achieved when the thickness ratio between the SiO_2 and the SiON layer is in the order of 0.2–0.3. The corresponding charge-to-breakdown data as determined from a Weibull plot are shown in Fig. 6.23, and indeed a maximum around this thickness ratio is observed. If the applied voltage is a major concern for an application, thickness ratios up to 0.67 also exhibit excellent Q_{BD} values. Another speculative idea could be the replacement of SiON by other high-k materials which would further minimize the voltage drop across the dielectric protection layer [142]. In order to quantify the effect of an additional SiON layer, comprehensive time-to-breakdown measurements were performed with unimplanted MOSLEDs having a 100 nm thick SiO_2 layer. The results were already given in Fig. 6.2 demonstrating that the use of an additional SiON layer increases the t_{BD} and thus the Q_{BD} value by more than two orders of magnitude.

The presence of a SiON layer cannot prevent the degradation processes in the bulk of the SiO_2 layer, especially the defect generation by band-to-band and trap-to-band impact ionization. However, it can diminish degradation processes at the gate-sided SiO_2 interface where electrons can release their kinetic energy plus the band offset energy after they were accelerated across the SiO_2 layer. The band offset between SiO_2 and SiON of ∼0.5 eV is much smaller than that between SiO_2 and

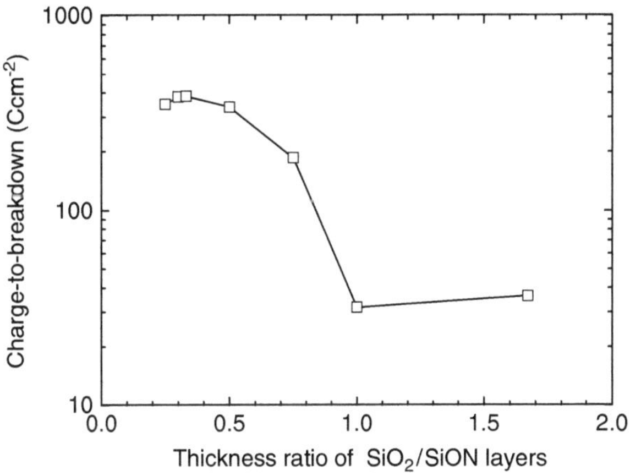

Fig. 6.23 Q_{BD} values as determined from a Weibull plot for Tb-implanted MOSLEDs having different SiO_2 and different SiON thicknesses as a function of their thickness ratio. The uncertainty of the Q_{BD} values amounts to a factor of 2–3

a metal which is usually >3 eV (see Fig. 3.1 for details). As a result, the fraction of electrons that can release an energy which is enough to cause damage at the corresponding SiO_2 interface is smaller in case of an additional SiON layer. Moreover, electrons entering the SiON layer face a massive deceleration because of (1) additional scattering due to the much higher defect density compared to SiO_2 and (2) the lower electric field due to the higher dielectric constant. Therefore, most of the electrons approaching the gate electrode are harmless with respect to their kinetic energy which is in good correspondence with the low number of hot electrons measured in SiON (Fig. 5.7).

In summary, MOS structures with an additional SiON layer degrade mainly by band-to-band and trap-to-band impact ionization in combination with charge trapping, whereas hole injection from the SiON layer seems to have a less deleterious effect than hole injection directly from the anode. Hydrogen release may play a major role in cases where the hydrogen introduced by the PECVD process is not extensively removed by an appropriate post-deposition annealing. In fact, reduced charge-to-breakdown values were measured for RE-implanted and unannealed MOSLEDs (Fig. 6.3), but already a moderate annealing drives out most of the hydrogen and weakens this degradation mechanism.

6.4 Potassium Codoping

As the EL quenching is closely linked to electron trapping in the vicinity of RE^{3+} LCs, charge compensation could be one possibility to alleviate this problem. However, injected holes and mobile hydrogen also cause degradation which normally overcompensate the advantages of charge neutralization. It is known from MOSFET

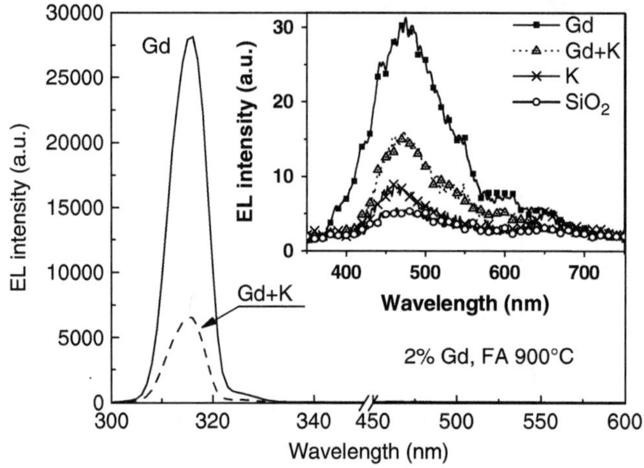

Fig. 6.24 EL spectra of Gd-implanted MOSLEDs without (*solid line*) and with a K co-implantation (*dashed line*). The *inset* shows a magnification of the weak EL in the visible for a Gd-, K-, Gd- and K-implanted MOSLEDs in comparison with unimplanted SiO_2 (after [139])

technology that the contamination with alkali ions is undesirable mostly because of the change of the threshold voltage and other device parameters during operation. In fact, Na co-implantation into RE-implanted MOSLEDs results in a dramatic deterioration of efficiency and stability. However, for a potassium co-implantation, the situation is different and more complex.

Figure 6.24 compares the EL spectra of MOSLEDs implanted with 2% Gd co-implanted and not co-implanted with 3% K at an injection current of $0.016\,\mathrm{A\,cm^{-2}}$. Most apparently, the EL intensity of the 316 nm line originating from Gd^{3+} decreases by a factor of 4 by the co-implantation. As shown in the inset of Fig. 6.24, the weak EL in the visible is also decreased, but only by a factor of approximately 2. This loss in efficiency is compensated by a gain of stability as demonstrated in Fig. 6.25. The change of the applied voltage under constant current injection (Fig. 6.25a) exhibits two remarkable features: At first, with increasing potassium concentration, the voltage required to achieve a certain injection current decreases significantly. Second, the electrical degradation in form of electron trapping is strongly delayed. This behaviour is mirrored for the ELI characteristic in Fig. 6.25b showing both a decrease of EL intensity and a delay of the EL quenching with increasing potassium concentration. In case of 3% K, the devices show an extremely long period (up to 10^{20}–$10^{21}\,\mathrm{e^-\,cm^{-2}}$) of stability with nearly constant applied voltages and EL intensities.

Generally, the implantation of alkali ions into SiO_2 leads to the formation of L-centres $\equiv$Si–$O^{(-)}$–$M^{(+)}$ with M as an alkali atom like Na or K. The bonding in the ground state of the L-centre is ionic in nature with a negative and positive charge centroid located closer to the oxygen and the alkali ion, respectively. In contrast, the first excited state of the L-centre exhibits a more covalent nature of the

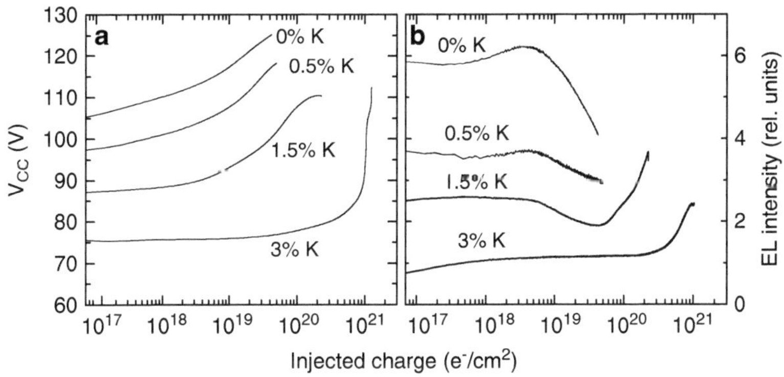

Fig. 6.25 Applied voltage (**a**) and the EL intensity (**b**) as a function of the injected charge under constant current injection of 0.016 A cm^{-2} for a MOSLED implanted with 2% Gd and co-implanted with different potassium concentrations (original data from [139])

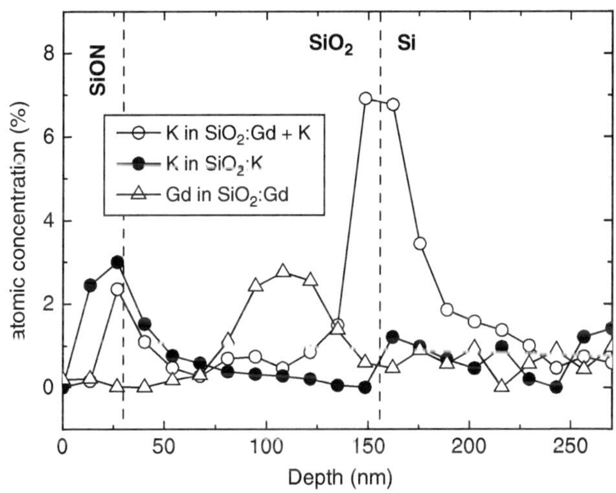

Fig. 6.26 K and Gd depth profiles in a SiON/SiO$_2$/Si structure obtained by means of AES. The samples contain a nominal peak concentration of 2% Gd and 3% K. K has been redistributed during a FLA treatment (after [140])

O–M bond [353]. The radiative transition from the first excited state of the L-centre to the ground state gives rise to a blue luminescence line, for which reason the weak blue EL in the inset of Fig. 6.24 is probably composed of that line and a contribution from ODCs created during implantation and annealing.

During annealing, K^+ diffuses towards the interfaces of the SiO$_2$ layer as deduced from AES measurements (Fig. 6.26). This results in an accumulation of positive charges at the Si–SiO$_2$ interface which in turn lowers the injection barrier for electrons and is responsible for the lowering of the applied voltage with increasing potassium concentration (Fig. 6.25a). As a consequence, the applied electric field across the SiO_2 layer and thus the average energy of the hot electrons are lower in case of high potassium concentrations. As already discussed before, this lowers the EL intensity but decreases the rate of defect creation during charge injection at the same time. Thus, the lowering of the injection barrier qualitatively explains the concurrent decrease of EL efficiency and increase of stability.

However, the effect of potassium co-implantation is not limited to that scenario. Under charge injection, several processes take place:

1. Release of K from the L-centre by impact ionization of hot electrons and migration of K^+ towards the Si–SiO_2 interface due to the applied electric field.
2. Trapping of electrons in defects in the close vicinity of RE^{3+} ions. This is identical with the EL quenching mechanism described in Sect. 6.2.2.
3. Neutralization of trapped electrons by K^+. The effects of this process are similar to the reactivation of blocked electrons by moderate temperature treatments as described in Sect. 6.2.3.

As potassium is predominantly located at both interfaces of the SiO_2 layer, process (1) mainly applies to potassium at the SiO_2–SiON interface because of the absence of hot electrons at the Si–SiO_2 interface. This leads to a continuous release of K^+ from L-centres located at the SiO_2–SiON interface with a subsequent migration towards the Si–SiO_2 interface. In this way, K^+ can neutralize electrons (process 3) that were previously trapped at defect sites close to RE^{3+} ions (process 2). Furthermore, electrons are also trapped at the Si–SiO_2 interface reducing the favourable effect of a low injection barrier. However, if the supply of potassium is high enough, the migrating K^+ ions approaching the Si–SiO_2 interface can compensate these electrons resulting in a long period where the effective injection barrier can be kept low. This is most probably the case of a potassium co-implantation of 3%.

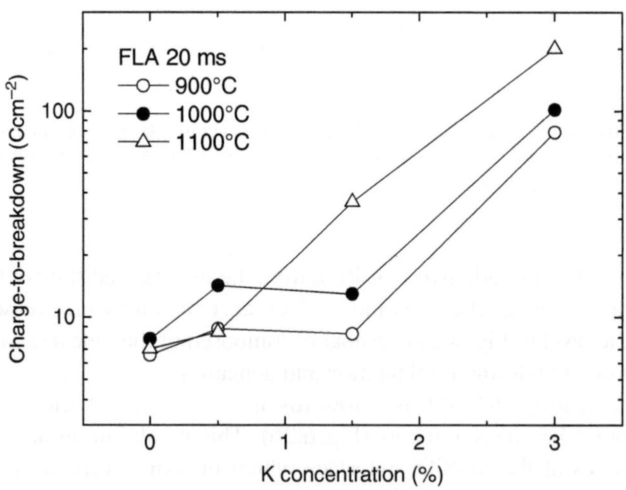

Fig. 6.27 Charge-to-breakdown values for Gd-implanted MOSLEDs as a function of K concentration for different thermal treatments (original data from [139])

The effect of the stability enhancement by K^+ is illustrated in Fig. 6.27 showing Q_{BD} values for Gd-implanted MOSLEDs as a function of potassium concentration for different thermal treatments. The increase of the Q_{BD} value for a concentration of 3% K by more than one order of magnitude compared to devices not implanted by K is clearly seen. However, this improvement has to be balanced with the decrease of the EL intensity in Fig. 6.25 for 3% K, and the question whether the K co-implantation is useful or not depends on the needs of potential applications.

Chapter 7
Applications

This short chapter intends to give a brief outlook on potential applications of RE-based MOS structures. Whereas Sect. 7.1 discusses the suitability of Si-based light emitters in general for a couple of applications, the focus of the subsequent chapter is on smart biosensing, which appears to be the most promising application at present.

7.1 Requirements for Si-Based Light Emitters

Today, there is a large palette of different approaches for Si-based light emission. All these approaches have different advantages and shortcomings, for which reason the suitability of a specific light emitter for a potential application depends strongly on the needs of this application. The key parameters that are of special interest for most of the applications are power efficiency, operation lifetime and operation voltage. In the past, the maximum switching speed was also a major concern, but today it is mostly assumed that a light emitter in combination with a fast modulator can work as well. At present, none of the various types of Si-based light emitters can compete with organic LEDs or III–V emitters in terms of these key parameters. However, Si-based light emitters have the advantage of integrability into Si chips, which can compensate their current shortcomings.

One of the main driving forces of Si-based light emission is the idea to replace electronic data processing by optical data processing in the framework of the ongoing miniaturization according to Moore's Law. Unfortunately, the requirements for an electrically driven light source integrated on chip and intended for chip-to-chip and intra-chip communication are really high. For such applications, power dissipation, and therefore, the power efficiency of integrated light emitters strongly matter, and the operation voltage should not deviate too much from the core voltage of present microelectronic circuits. Moreover, an operation lifetime of several thousands of hours is mandatory. Thus, electrically driven, *integrated* light emitters are not available at present. In addition, the question whether Si-based light emitters will be applied in future for this type of application depends not only on the future progress in the field of Si-based light emission, but also on the development of alternative technologies. Current solutions use hybrid techniques to mount a suitable light source (e.g. a III–V emitter) on a chip and to couple their light into appropriate waveguides. If this technology becomes more smart and cost-effective in future, it will be a serious competitor for Si-based light emission. Moreover, modern concepts of integrated photonic circuits try to minimize the number of required light emitters with one central light source feeding the whole chip or even several chips in optimum [354]. This weakens the need for miniaturization for such kind of light sources, and thus, the chances of Si-based light emission in this field.

Fortunately, there are further applications whose requirements are less severe. In case of an integrated optocoupler, which separates a high voltage from a low voltage circuitry, the higher operation voltage could be even an advantage. In [99], such an optocoupler composed of a Ge-implanted MOSLED and a pin diode processed in amorphous Si was successfully demonstrated. This type of optocoupler exhibited a linear transfer characteristic over more than three orders of magnitude and has the advantage that the high voltage level can be tuned, within certain limits, by varying the SiO_2 layer thickness of the light emitter. The design can be easily applied to RE-implanted MOSLEDs.

Another very promising application is biosensing in which the usual external light source is replaced by an integrated one. In most cases, only a few light sources with an electrical power requirement in the mW range are needed, which is why power dissipation and thus the power efficiency are uncritical. There is also no reason for not tolerating higher voltages; at least, if they are below the line voltage. Whether the operation lifetime is a major concern or not depends on the type of application: if intended to be used in continuous monitoring tasks, it probably matters; if used for disposable, it should not be critical at all.

However, in order to become a serious competitor to classical analysis methods, Si-based light emitters must offer an essential advantage. Undoubtedly, this is the enormous potential for miniaturization. Current standard methods for the detection of small amounts of organic substances in water are gas and liquid chromatography which usually enrich the sample by up to three orders of magnitude. The technique of enzyme linked immunosorbent assay (ELISA) is commonly applied as a sensitive technique for the detection of organic substances in the parts per billion range without sample preconcentration [355]. However, all these methods are normally used in standard laboratories, and it is difficult to automate them or to use it for point-of-care measurements. The first approach for portable systems came with the development of the river analyzer (RIANA) [356] and the AWACSS (automated water analyzer computer supported system) system [357] which already take advantage of small-sized waveguides. Both of them are based on the total internal reflection fluorescence method which, however, uses external laser diodes as light sources. Although these systems are designed for an automated field use, their mobility is still limited by their size.

The high degree of miniaturization offered by Si-based light emitters results in an enormous device shrinking with the corresponding savings of material and energy, and in some cases, only integrated light sources allow the usage of point-of-care systems. Moreover, due to its compatibility to current Si technology, Si-based light emitters can fully take advantage from the economy of scale. In addition, the shape and design of Si-based light sources are only limited by the possibilities of modern lithography and process technology, for which reason customized solutions are very easy to realize. Finally, such types of biosensors (Sects. 7.2.3 and 7.2.4) can be regarded as a platform technology which could be applied to the fabrication of a large class of biosensor systems just by changing the recognition element. Nevertheless, at first, the current state of Si-based biosensing is briefly described in the following two chapters, before future concepts will be discussed.

7.2 Si-Based Optical Biosensing

7.2.1 Introduction

In general, a biosensor is composed of a biorecognition element, a transducer and a detection unit. The biorecognition element consists of an ensemble of appropriate biomolecules on a bioactive surface and has the task to trigger a specific reaction if an analyte is present. In most cases, this specific reaction is the selective immobilization or capture of the analyte on the surface. Depending on the type of analyte, the biomolecule can be a receptor, an antibody, an enzyme or something different. The transducer transforms the result of this specific reaction into a physical or chemical signal which can finally be detected. This basic scheme is sketched in Fig. 7.1a and can be found with slight variations in most of the textbooks about biosensing (e.g. [358–360]). In case of optical biosensing, the presence of an analyte changes an optical property of the transducer which is probed by the light from a light source (Fig. 7.1b). In most cases, the specific reaction of the biorecognition element changes the effective refraction index and/or the effective optical thickness

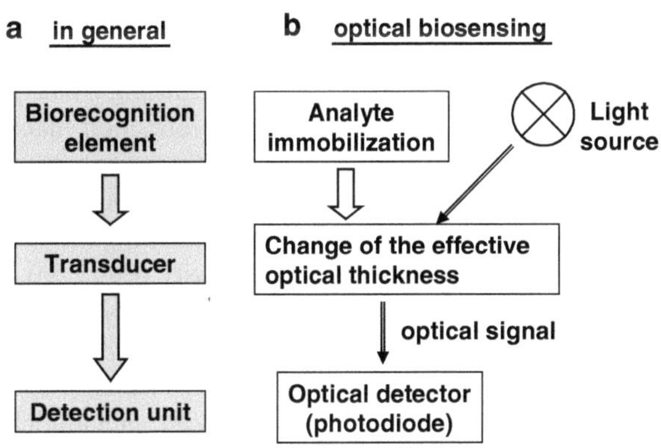

Fig. 7.1 Basic scheme of a biosensor (**a**) and a more specific example for optical biosensing (**b**)

of the transducer. However, the optical signal can also be the light emission of a fluorescence marker.

At present, there are no ready-to-use biosensors which take advantage of an integrated light emitter. One principle in doing so, namely, the direct fluorescence analysis, was proposed in [361] and partly realized in a non-Si-based approach in [362]. In the latter case, integrated, organic LEDs arranged in arrays were used to excite an oxygen-sensitive dye whose efficiency and PL decay time change with the oxygen concentration. Glucose, lactate and ethanol are oxidized with the help of the corresponding enzymes, and the consumption of oxygen is probed by the dye. However, Si-based materials have been used as transducers in combination with external light sources. Thus, the next chapter gives a short review of Si-based materials as passive transducers, whereas Sect. 7.2.3 discusses the direct fluorescence analysis at the example of RE-implanted MOSLEDs. Finally, Sect. 7.2.4 addresses more advanced concepts, namely photonic circuits with different waveguide architectures as transducer.

7.2.2 Si-Based Materials as Passive Transducers

In the last 15 years, optical biosensing using Si-based materials became popular in the scientific community. Among the various Si-based materials, porous Si has possibly attracted the highest interest for biosensing because of its large internal surface. This internal surface, which can range up to 100 $m^2\,cm^{-3}$ and more, allows the immobilization of a much higher number of analyte molecules than planar surfaces. Indeed, an increase of sensitivity by a factor of 9, measured as the angular resonance shift of the reflectance spectrum, was found in porous SiO_2 compared to bulky SiO_2 waveguide sensors [363]. Encouraging results were also obtained from porous Al_2O_3 [364] and TiO_2 [365], and in [366] the hypothetical sensitivity of a porous Si membrane guide as a function of porosity is calculated.

The effective refraction index of porous Si is composed of that of Si and that of the medium filling the pores (usually air or water). Biomolecules which are immobilized in the pores displace the filler medium causing a change of the pore's refraction index and thus a change of the effective refraction index of porous Si. This change is usually probed by measuring the reflectance as a function of the incident angle or the wavelength. The angular reflectance is usually measured in the so-called Kretschmann configuration [367] in which a polarized light beam is coupled in a waveguide structure or Fabry–Pérot cavity (Fig. 7.2a). The change of the effective refraction index of the porous Si layer caused by the immobilization of biomolecules leads to a shift of the reflectance minimum which is proportional to the biomolecule load in the ideal case. For example, this technique was successfully applied for the sensing of DNA oligonucleotides [368], glutaraldehyde [369] and for the monitoring of cell proliferation on a waveguide surface [370].

In a so-called interferometric measurement, the reflectance spectra of porous Si with and without biomolecules are compared with each other (Fig. 7.2b). Analogous

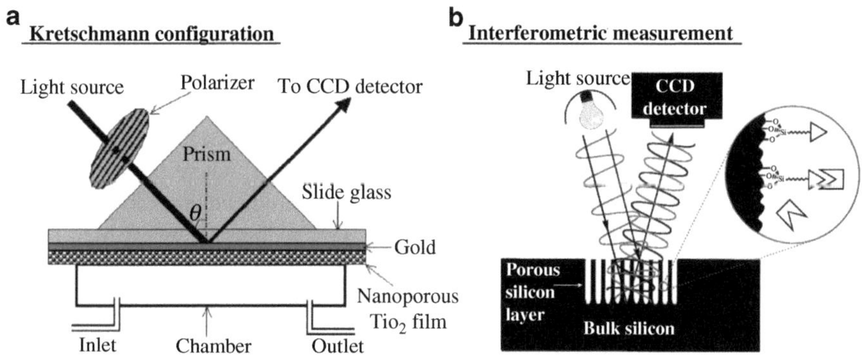

Fig. 7.2 Basic scheme of a Kretschmann configuration (**a**) after [365] and example of a simple interferometric measurement without a reference arm (**b**) after [372]

to the angular dependence, the reflectance spectrum on the wavelength scale exhibits interference patterns which shift with the biomolecule load. The change of the effective refraction index can be induced from the difference spectrum or, for more complex cases, the fast Fourier transform of the spectra [371]. Interferometric measurements were used e.g. to trace DNA oligonucleotides [372, 373], bovine serum albumin [371, 374], horseradish peroxidase [375], the binding of streptavidin [376] or simply the refraction index of liquids [377] or gases [378] filling the pores of porous Si. Both methods, which are based on the analysis of the reflectance spectrum, are close to the direct ellipsometric determination of the refraction index as done e.g. in [379].

The sensitivity of a potential biosensor can be enhanced by using more advanced transducer geometrics. In [371], two porous Si layers with small and large pores were applied which allow an additional separation of the biomolecules according to their size. Bragg mirrors made of alternating porous Si layers with high and low porosity [369, 373, 380, 381] sharpen the resonance in the reflectance spectrum, and wavelength shifts can be determined more precisely. For further optimization, the Bragg mirror can also be designed as a rugate filter (with a sinusoidal refraction index profile) [382–384].

Other measurement techniques exploit the PL properties of porous Si processed as a microcavity. This PL spectrum also exhibits sharp resonance lines due to the microcavity which shift with increasing concentration of the immobilized analyte [385, 386]. In [387], myoglobin was traced by analysing the PL quenching of porous Si caused by the dehydrogenization of the porous Si surface during the binding reaction of the analyte and the subsequent capture by the immune complex. Porous Si was also used in a chemiluminescence assay in order to detect *E. coli* [388].

Besides porous Si, other Si-based materials have also been utilized for biosensing. Au nanoparticles were deposited on a SiO_2 surface [389, 390] in order to take advantage from the surface plasmon resonance which shifts with binding events on the Au surface. More details to this method can be found in [391].

Recently, photonic crystals, preferably processed on SOI (silicon on insulator) wafers, have been used in combination with suitable waveguides for the detection of biomolecules [392, 393] and other chemical substances [394]. In this case, the presence of biomolecules immobilized in the cavities of the photonic crystal will shift the resonance wavelength of the crystal. Another approach is the use of Si ring and disc resonators ([395, 396] and references therein) where the immobilization of biomolecules in such resonator cavities also strongly changes the resonance frequency of the system. Further details about optical biosensing with Si-based materials can be found in the following reviews [396–399].

7.2.3 The Concept of Direct Fluorescence Analysis

One concept of using integrated light emitters for biosensing is direct fluorescence analysis in which the light emitter is directly placed beneath the sample. In the following, this concept is discussed at a specific example using RE- and Ge-implanted MOSLEDs for the detection of oestrogen in waterish solutions. The basic module is a chip with one or several RE-implanted MOSLEDs which are passivated with a thick SiO_2 layer deposited by PECVD in order to prevent shortcuts in fluid media. On top of the passivation layer a bioactive film is processed which is composed of a linker and a receptor. The linker bounds the receptor on the SiO_2 surface and is realized by a specific silanization procedure to avoid chemical processing steps which might be harmful to the light emitters [400]. The receptor should be selective to the analyte and has the task to immobilize it. A basic scheme with the light emitter, the linker, the receptor and a dye-labelled analyte is given in Fig. 7.3. For the actual measurement, the MOSLED emits a primary luminescence which excites the dye. The dye in turn emits a secondary luminescence at a longer wavelength which passes an edge filter blocking the primary light signal. In the ideal case, the detector records only the secondary signal which is a measure for the concentration of the dye-labelled analyte.

The measurement sequence as described in [401, 402] can be divided into three phases and starts with a light emitter chip provided with the aforementioned bioactive film with the linker and the receptor (Fig. 7.4a). In phase (1), the waterish sample passes the surface of the light emitter via a suitable microfluidic, and analyte molecules being solved in the sample are immobilized at the receptors with a certain probability (Fig. 7.4b). Depending on the concentration of the analyte, a fraction p of the receptor sites is occupied. In phase (2), a reference solution containing dye-labelled analytes passes the light emitter surface after a suitable washing step (Fig. 7.4c). This step has the task to occupy all remaining free receptor sites with dye-labelled analytes. The advantage is that the reference solution can be prepared in advance and that no chemical reactions must be run involving the sample which normally has an unknown composition. Phase (3) is the optical measurement as described above (Fig. 7.4d). The result is a signal proportional to the fraction $1 - p$ of the dye-labelled analyte allowing the calculation of the original analyte

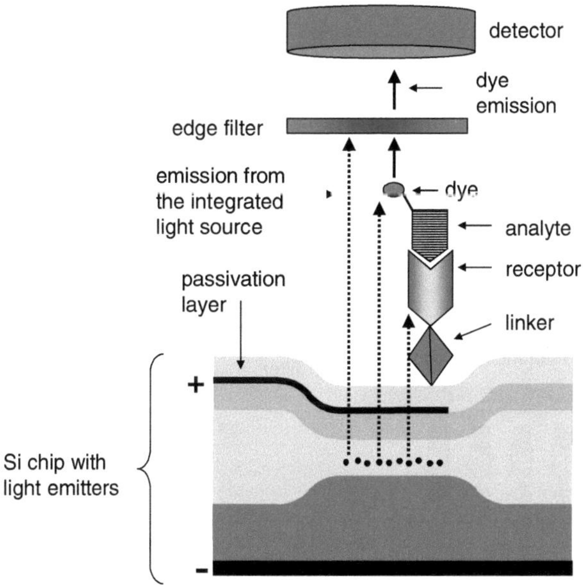

Fig. 7.3 Basic scheme of a biosensor using direct fluorescence analysis with RE-implanted MOSLEDs. The main components are the integrated light emitter (*lower part*), the bioactive layer (*medium part*) and the signal detection (*upper part*)

concentration in the sample. Thus, low and high analyte concentrations will cause a high and low detector signal, respectively.

The principle was already proven for the specific case of oestrogen in water [403]. One of the issues, which is interesting from an optical point of view, is the interaction between the integrated light emitter and the dyes used to label the analyte. Figure 7.5 displays the EL spectra of a Gd-, Ge- or Tb-implanted MOSLED with and without a 3 μmol solution containing the dye Qdot® 800 [404] on top. The dye whose emission wavelength is about 800 nm is composed of a nanometre-sized crystal of CdSeTd as core, a ZnS shell around it and a final polymer coating which results in a broadband absorption in the green-blue and UV with increasing extinction coefficient at decreasing wavelength [404]. Consequently, the excitation with the 316 nm line from Gd^{3+} and the blue emission from Ge-implanted devices leads to a strong dye emission at 800 nm (Fig. 7.5a, b). Although the dye signal is very strong, it has to be considered that possibly not only the dye can show a luminescence under short-wavelength irradiation. If excited mainly at 540 nm by the Tb-based MOSLED, the dye signal is less intense (Fig. 7.5c), but nevertheless in a spectral region with a low background. Therefore, all three light emitters seem to be suitable for the use in biosensing, but the question which light emitter will be the best depends on the signal-to-noise ratio.

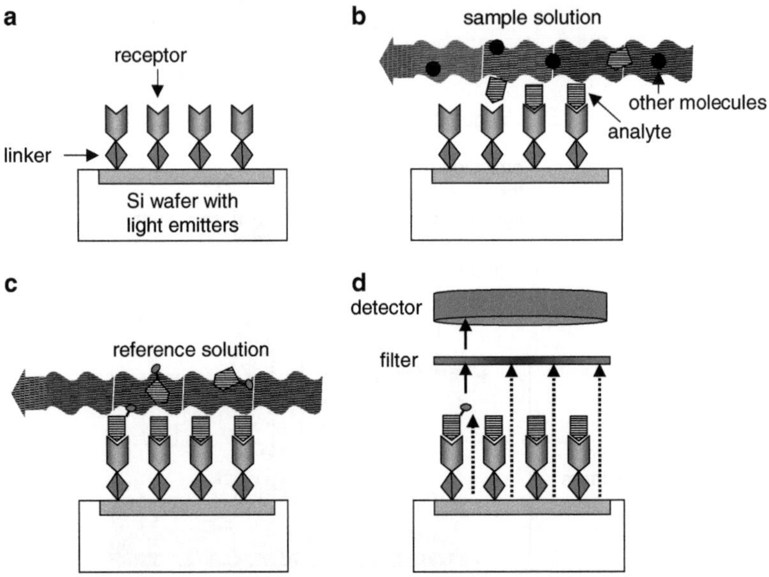

Fig. 7.4 Basic scheme of the measurement sequence: Starting from (**a**), a sample solution passes the functionalized surface which immobilizes the analyte (**b**). Afterwards, a reference solution containing dye-labelled analyte molecules passes the functionalized surface in order to occupy all receptor places which are still unoccupied (**c**). Finally, the optical measurement determines the amount of dye-labelled analytes and thus the number of analytes originally bound to the receptors

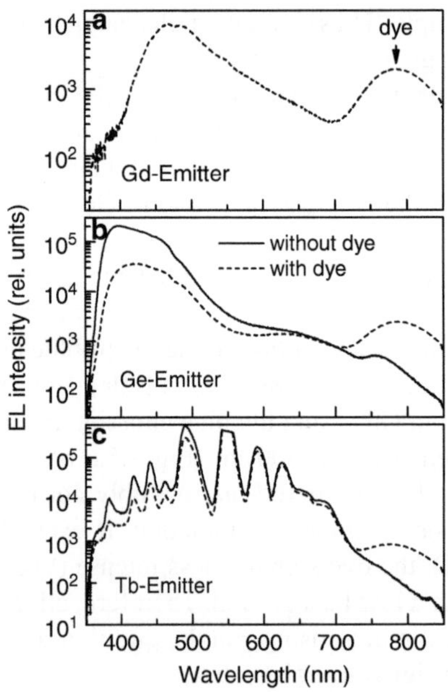

Fig. 7.5 EL spectra of Gd-, Ge- and Tb-implanted MOSLEDs with and without a 3 μmol waterish solution containing the dye QDot® 800 [404] on top

7.2.4 *Biosensing with Si-Based, Integrated Photonic Circuits*

All the aforementioned reports (Sect. 7.2.2) about Si-based materials for biosensing use external light sources and external detectors which in general limit the potential for miniaturization. Also the concept of direct fluorescence analysis still uses an external detector. Thus, the next logical step on the road of miniaturization is the integration of the detector leading to a so-called photonic circuit. Such a circuit comprises the light emitter, the detector and a waveguide which connect both components and serves as transducer for the biosensor.

Except light emitters, there are already a couple of satisfying solutions for Si-based waveguides, modulators and detectors. It is not in the focus of this book to discuss all these approaches; more details can be easily found in [4, 6]. However, up to now the different attempts to combine all these components together rarely went beyond the integration of one component with a waveguide. Most common is the combination of a waveguide with an integrated detector, as e.g. demonstrated for integrated Si [1, 4, 405, 406] and Ge detectors [407, 408]. Recently, this combination has been successfully adapted for the detection of bovine serum albumin [409]. The number of studies reporting about photonic circuits with integrated light emitters is very low. First experimental approaches place light emitting structures in a waveguide [40, 43, 410] or a photonic crystal [270, 273] and pump them optically. In order to obtain a photonic circuit with *electrically driven* light emitters, bonding techniques have been used to link vertical cavity surface emitting lasers (VCSELs) with CMOS detection elements [411]. Some endeavours have been made to couple the light from integrated organic LEDs into a waveguide [412] and detected it with organic photodiodes [413, 414]. To the best of our knowledge, there is only one report about a photonic circuit with an *electrically driven* light emitter in Si technology [415, 416], in which the light of a Si p–i–n diode is transmitted to a $Si/Si_{1-x}Ge_x/Si$ diode via a Si_3N_4 waveguide. For RE-implanted MOSLEDs processed on SOI wafers, a concept is known to couple its light into a waveguide [417].

Figure 7.6 is a snapshot how future photonic circuits could work as smart and advanced biosensors. All the ideas already exist, but mostly in the context with external light sources or other measurement principles. In general, the light source emits light which is coupled into the waveguide. When the light wave propagates through the waveguide, it can be manipulated in various ways depending on the binding events of the recognition element. This manipulation is measured by the integrated detector as a change of intensity.

Starting the discussion with Fig. 7.6a, the evanescent field of the propagating light wave leaks out of the waveguide and may excite dyes which are attached either to the analyte or to the receptor (if the quenching of PL is a measure for the presence of an analyte). Consequently, the emission of the dye is the measurement signal. This concept is derived from the waveguide sensor presented in [356, 357], but it is rather an intermediate state than a fully integrated photonic circuit. In fact, it still has two drawbacks: it uses dyes as fluorescence markers and the signal has to be detected by an external detector. Eliminating these two disadvantages leads

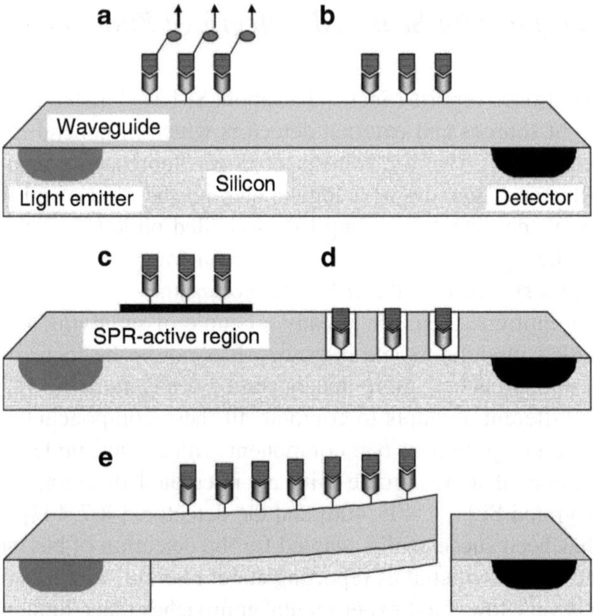

Fig. 7.6 Possible concepts of optical biosensors based on fully integrated photonic circuits: Excitation of fluorescence markers by the evanescent field (**a**), Absorption of the evanescent field (**b**), Change of the waveguide transmission due to a shift of the surface plasmon resonance (**c**) or due to photonic crystals (**d**) and use of a cantilever structure (**e**)

to Fig. 7.6b in which the propagation of the light wave is more or less disturbed depending on how strong the evanescent field is absorbed. This method requires that the absorption of the analyte or the analyte–receptor complex is high enough and differs strongly from the case if no analyte is present.

The interaction of the light wave with the analyte can be enhanced by utilizing the surface plasmon resonance effect (Fig. 7.6c). For doing so, the waveguide is cladded in the active region by a thin metal layer, preferably gold. Depending on the refraction index of the analyte, the transmission through the waveguide shifts considerably on the wavelength scale [418]. Another method is the incorporation of cavities in the waveguide which starts with a simple gap and ends up with two-dimensional photonic crystals (Fig. 7.6d). As already mentioned, the presence of immobilized biomolecules in the cavities will shift the resonance wavelength and thus the transmission of the waveguide. Finally, Fig. 7.6e was inspired by an idea presented in [419]. In this concept, the waveguide is interrupted, and one side is implemented as a cantilever fabricated in Si technology. The immobilization of biomolecules on one side of the cantilever can cause a change of the surface stress of the active side with respect to the other, inactive side of the cantilever. In the original concept, the resulting bending of the cantilever in the nanometre range is detected by the position change of a laser beam which is reflected from the cantilever surface. In the context of photonic circuits, the cantilever can additionally be equipped with a waveguide, and the mismatch between the cantilever waveguide and the adjacent waveguide leading to the integrated detector can considerably change the transmission through the whole system.

Of course, the collection of aforementioned concepts of photonic circuits for biosensing is far from being complete but gives a good impression of the potential of photonic circuits for biosensing. Thus, there is a really high probability that this type of biosensor will come, albeit the specific implementation is, clearly, not yet known. Such photonic circuits may comprise RE-implanted MOSLEDs, another type of Si-based light emitters, or light emitters which do not base on Si, if the used hybrid technology is sophisticated enough. It might be even possible that integrated photonic circuits which do not base on Si at all will dominate the future field of biosensors, e.g. if plastic electronics will see dramatic advances in the next future.

References

1. H. Zimmermann, *Integrated Silicon Optoelectronics* (Springer, Berlin, 2000)
2. L. Pavesi, S. Gaponenko, L. Dal Negro (eds.), *Towards the First Silicon Laser*, NATO Science Series (Kluwer, Dordrecht, 2002)
3. S. Ossicini, L. Pavesi, F. Priolo, *Light Emitting Silicon for Microphotonics* (Springer, Berlin, 2003)
4. L. Pavesi, D.J. Lockwood (eds.), *Silicon Photonics* (Springer, Berlin, 2004)
5. G.T. Reed, A.P. Knights, *Silicon Photonics: An Introduction* (Wiley, Chichester, 2004)
6. H. Zimmermann, *Silicon Optoelectronic Integrated Circuits* (Springer, Berlin, 2004)
7. L. Pavesi, G. Guillot (eds.), *Optical Interconnects. The Silicon Approach* (Springer, Berlin, 2006)
8. G.T. Reed, A.P. Knights, *Silicon Photonics: The State of the Art* (Wiley, Chichester, 2008)
9. L.T. Canham, Appl. Phys. Lett. **57**, 1046 (1990)
10. J.E. Lilienfeld, US patent No. 1745175A, 1925
11. G. Ekspong (ed.), *Nobel Lectures, Physics 1996–2000* (World Scientific, Singapore, 2002), p. 474
12. F. Szabadvary, *Two-Hundred-Year Impact of Rare Earths on Science* (Elsevier, New York, 1988), p. 33
13. C.T. Sah, Proc. IEEE **76**, 1280 (1988)
14. R.B. Fair, Proc. IEEE **86**, 111 (1998)
15. D. Kahng, US Patent No. 3102230A, 1963
16. J.W. Allen, H.G. Grimmeiss, Mat. Sci. Forum **590**, 1 (2008)
17. H.J. Round, Electrical World 19 (1907) 309 [Reprint in *Semiconductor Devices: Pioneering Papers*, S.M. Sze (ed.), World Scientific, Singapore, 1991, 879]
18. O.V. Lossev, Philos. Mag. Ser. **6**, 1024 (1928)
19. J.R. Haynes, H.B. Briggs, Phys. Rev. **86**, 647 (1952)
20. J.R. Haynes, W.C. Westphal, Phys. Rev. **101**, 1676 (1956)
21. H. Ennen, J. Schneider, G. Pomrenke, A. Axmann, Appl. Phys. Lett. **43**, 943 (1983)
22. T. Oesterreich, C. Swiatkowski, I. Broser, Appl. Phys. Lett. **56**, 446 (1990)
23. A. Polman, A. Lidgard, D.C. Jacobson, P.C. Becker, R.C. Kistler, G.E. Blonder, J.M. Poate, J. Appl. Phys. Lett. **57**, 2859 (1990)
24. T. Shimizu-Iwayama, M. Ohshima, T. Niimi, S. Nakao, K. Saitoh, T. Fujita, N. Itoh: J. Phys. Condens. Matt. **5**, L375 (1993)
25. H.A. Atwater, K.V. Shcheglov, S.S. Wong, K.J. Vahala, R.C. Flagan, M.L. Brongersma, A. Polman, Mater. Res. Soc. Symp. Proc. **316**, 409 (1994)
26. W. Skorupa, R.A. Yankov, I.E. Tyschenko, H. Fröb, T. Böhme, K. Leo, Appl. Phys. Lett. **68**, 2410 (1996)
27. Y. Maeda, N. Tsukamoto, Y. Yazawa, Y. Kanemitsu, Y. Masumoto, Appl. Phys. Lett. **59**, 3168 (1991)
28. L. Skuja, J. non-cryst. solids **149**, 77 (1992)

29. C.M. Yang, K.V. Shcheglov, K.J. Vahala, H.A. Atwater, Nucl. Instrum. Methods B **106**, 433 (1995)
30. C.A. Dimitriadis, J.H. Werner, S. Logothetidis, M. Stutzmann, J. Weber, R. Nesper, J. Appl. Phys. **68**, 1726 (1990)
31. D. Leong, M. Harry, K.J. Reeson, K.P. Homewood, Nature **387**, 686 (1997)
32. V. Kveder, M. Badylevich, E. Steinman, A. Izotov, M. Seibt, W. Schröter, Appl. Phys. Lett. **84**, 2106 (2004)
33. Z.H. Lu, D.J. Lockwood, J.-M. Baribeau, Nature **378**, 258 (1995)
34. G. Abstreiter, K. Eberl, E. Friess, W. Wegscheider, R. Zachai, J. Cryst. Growth **95**, 431 (1989)
35. J. Engvall, J. Olajos, H.G. Grimmeiss, H. Presting, H. Kibbel, E. Kasper, Appl. Phys. Lett. **63**, 491 (1993)
36. S. Wang, A. Eckau, E. Neufeld, R. Carius, Ch. Buchal, Appl. Phys. Lett. **71**, 2824 (1997)
37. E.Ö. Sveinbjörnsson, J. Weber, Appl. Phys. Lett. **69**, 2686 (1996)
38. W.L. Ng, M.A. Lourenço, R.M. Gwilliam, S. Ledain, G. Shao, K.P. Homewood, Nature **410**, 192 (2001)
39. J.M. Sun, T. Dekorsy, W. Skorupa, B. Schmidt, A. Mücklich, M. Helm, Phys. Rev. B **70**, 155316 (2004)
40. L. Pavesi, L. Dal Negro, C. Mazzoleni, G. Franzó, F. Priolo, Nature **408**, 440 (2000)
41. R. Claps, D. Dimitropoulos, V. Raghunathan, Y. Han, B. Jalali, Opt. Exp. **11**, 1731 (2003)
42. H. Rong, A. Liu, R. Jones, O. Cohen, D. Hak, R. Nicolaescu, A. Fang, M. Paniccia, Nature **433**, 292 (2005)
43. J. Liu, X. Sun, R. Camacho-Aguilera, L.C. Kimerling, J. Michel, Opt. Lett. **35**, 679 (2010)
44. N. Daldosso, L. Pavesi, Laser Photon. Rev. **3**, 508 (2009)
45. D.J. DiMaria, J.R. Kirtley, E.J. Pakulis, D.W. Dong, T.S. Kuan, F.L. Pesavento, T.N. Theis, J.A. Cutro, S.D. Brorson, J. Appl. Phys. **56**, 401 (1984)
46. P.D. Altukhov, G.V. Ivanov, E.G. Kuzminov, Solid State Electron. **42**, 1657 (1998)
47. M.-J. Chen, C.-F. Lin, M.H. Lee, S.T. Chang, C.W. Liu, Appl. Phys. Lett. **79**, 2264 (2001)
48. M.-J. Chen, C.-F. Lin, W.T. Liu, S.T. Chang, C.W. Liu, Appl. Phys. **89**, 323 (2001)
49. C.W. Liu, C.-H. Lin, M.H. Lee, S.T. Chang, Y.-H. Liu, M.-J. Chen, C.-F. Lin, Appl. Phys. Lett. **78**, 1397 (2001)
50. C.-H. Lin, F. Yuan, B.-C. Hsu, C.W. Liu, Solid State Electron. **47**, 1123 (2003)
51. H.-C. Lee, S.-C. Lee, Y.-P. Lin, C.-K. Liu, Jpn. J. Appl. Phys. **44**, 3867 (2005)
52. T.-M. Wang, C.-H. Chang, S.-J. Chang, J.-G. Hwua, J. Vac. Sci. Technol. A **24**, 2049 (2006)
53. C.-F. Lin, P.-F. Chung, M.-J. Chen, Opt. Lett. **27**, 713 (2002)
54. X. Yu, W. Seifert, O.F. Vyvenko, M. Kittler, T. Wilhelm, M. Reiche, Appl. Phys. Lett. **93**, 041108 (2008)
55. G.Z. Ran, J.S. Fu, W.C. Qin, B.R. Zhang, Y.P. Qiao, G.G. Qin, Thin Solid Films **388**, 213 (2001)
56. G.G. Qin, A.P. Li, B.R. Zhang, B.-C. Li, J. Appl. Phys. **78**, 2006 (1995)
57. G.G. Qin, A.P. Li, Y.X. Zhang, Phys. Rev. B **54**, R11122 (1996)
58. P.F. Trwoga, A.J. Kenyon, C.W. Pitt, J. Appl. Phys. **83**, 3789 (1998)
59. N. Lalic, J. Linnros, J. Luminesc. **80**, 263 (1999)
60. G. Franzò, A. Irrera, E.C. Moreira, M. Miritello, F. Iacona, D. Sanfilippo, G. Di Stefano, P.G. Fallica, F. Priolo, Appl. Phys. A **74**, 1 (2002)
61. M. Kulakci, U. Serincan, R. Turan, Semicond. Sci. Technol. **21**, 1527 (2006)
62. A. Irrera, F. Iacona, I. Crupi, C.D. Presti, G. Franzò, C. Bongiorno, D. Sanfilippo, G. Di Stefano, A. Piana, P.G. Fallica, A. Canino, F. Priolo, Nanotechnology **17**, 1428 (2006)
63. O. Jambois, B. Garrido, P. Pellegrino, J. Carreras, A. Pérez-Rodríguez, J. Montserrat, C. Bonafos, G. BenAssayag, S. Schamm, Appl. Phys. Lett. **89**, 253124 (2006)
64. M. Perálvarez, J. Carreras, J. Barreto, A. Morales, C. Domínguez, B. Garrido, Appl. Phys. Lett. **92**, 241104 (2008)
65. M. Perálvarez, J. Barreto, J. Carreras, A. Morales, D. Navarro-Urrios, Y. Lebour, C. Domínguez, B. Garrido, Nanotechnology **20**, 405201 (2009)
66. L. Rebohle, J. von Borany, R.A. Yankov, W. Skorupa, I.E.Tyschenko, H. Fröb, K. Leo, Appl. Phys. Lett. **71**, 2809 (1997)

67. L. Rebohle, J. von Borany, H. Fröb, W. Skorupa, Appl. Phys. B **71**, 131 (2000)
68. P. Knàpek, B. Rezek, D. Muller, J.J. Grob, R. Lèvy, K. Luterovà, J. Kocka, I. Pelant: Phys. Stat. Sol. A **167**, R5 (1998)
69. D. Muller, P. Knàpek, J. Fauré, B. Prevot, J.J. Grob, B. Hönerlage, I. Pelant, Nucl. Instr. Meth. B 148, 997 (1999)
70. A.G. Nassiopoulou, V. Ioannou-Sougleridis, P. Photopoulos, A. Travlos, V. Tsakiri, D. Papadimitriou, Phys. Stat. Sol. A **165**, 79 (1998)
71. G.-R. Lin, C.-J. Lin, J. Appl. Phys. **95**, 8484 (2004)
72. Y. Liu, T.P. Chen, L. Ding, M. Yang, J.I. Wong, C.Y. Ng, S.F. Yu, Z.X. Li, C. Yuen, F.R. Zhu, M.C. Tan, S. Fung, J. Appl. Phys. **101**, 104306 (2007)
73. L.-S. Liao, X.-M. Bao, N.-S. Li, X.-Q. Zheng, N.-B. Min, Solid State Commun. **97**, 1039 (1996)
74. H.Z. Song, X.M. Bao, N.S. Li, J.Y. Zhang, J. Appl. Phys. **82**, 4028 (1997)
75. B. Garrido, M. López, A. Pérez-Rodríguez, C. García, P. Pellegrino, R. Ferré, J.A. Moreno, J.R. Morante, C. Bonafos, M. Carrada, A. Claverie, J. de la Torre, A. Souifi, Nucl. Instrum. Methods B **216**, 213 (2004)
76. T. Matsuda, K. Nishihara, M. Kawabe, H. Iwata, S. Iwatsubo, T. Ohzone, Solid State Electron. **48**, 1933 (2004)
77. L. Ding, T.P. Chen, Y. Liu, M. Yang, J.I. Wong, K.Y. Liu, F.R. Zhu, S. Fung, Nanotechnology **18**, 455306 (2007)
78. T. Matsuda, Y. Honda, T. Ohzone, Solid State Electron. **42**, 129 (1998)
79. M. Perálvarez, C. García, M. López, B. Garrido, J. Barreto, C. Domínguez, J.A. Rodríguez, Appl. Phys. Lett. **89**, 051112 (2006)
80. J. Barreto, M. Perálvarez, J.A. Rodríguez, A. Morales, M. Riera, M. López, B. Garrido, L. Lechuga, C. Domínguez, Physica E **38**, 193 (2007)
81. A. Irrera, G. Franzò, F. Iacona, A. Canino, G. Di Stefano, D. Sanfilippo, A. Piana, P.G. Fallica, F. Priolo, Physica E **38**, 181 (2007)
82. J. Carreras, O. Jambois, S. Lombardo, B. Garrido, Nanotechnology **20**, 155201 (2009)
83. M. Sopinskyy, V. Khomchenko, Curr. Opin. Solid State Mater. Sci. **7**, 97 (2003)
84. G.G. Qin, S.Y. Ma, Z.C. Ma, W.H. Zong, Y. Li-ping, Solid State Commun. **106**, 329 (1998)
85. C.L. Heng, B.R. Zhang, Y.P. Qiao, Z.C. Ma, W.H. Zong, G.G. Qin, Physica B **270**, 104 (1999)
86. L. Heikkilä, T. Kuusela, H.-P. Hedman, Superlattices Microstruct. **26**, 157 (1999)
87. G. Pucker, P. Bellutti, M. Cazzanelli, Z. Gaburro, L. Pavesi, Opt. Mater. **17**, 27 (2001)
88. S.Y. Ma, Thin Solid Films **402**, 222 (2002)
89. B. Averboukh, R. Huber, K.W. Cheah, Y.R. Shen, G.G. Qin, Z.C. Ma, W.H. Zong, J. Appl. Phys. **92**, 3564 (2002)
90. A. Karsenty, A. Sa'ar, N. Ben-Yosef, J. Shappir, Appl. Phys. Lett. **82**, 4830 (2003)
91. J.H. Kaang, Y.D. Kim, K.M. Cha, H.J. Cheong, Y. Kim, J.-Y. Yi, H.J. Bark, T.H. Chung, J. Korean Phys. Soc. **45**, 1065 (2004)
92. D.Y. Chen, D.Y. Wei, J. Xu, P.G. Han, X. Wang, Z.Y. Ma, K.J. Chen, W.H. Shi, and Q.M. Wang, Semicond. Sci. Technol. **23**, 015013 (2008)
93. A. Anopchenko, A. Marconi, E. Moser, S. Prezioso, M. Wang, L. Pavesi, G. Pucker, P. Bellutti, J. Appl. Phys. **106**, 033104 (2009)
94. T. Wang, D.Y. Wei, H.C. Sun, Y. Liu, D.Y. Chen, G.R. Chen, J. Xu, W. Li, Z.Y. Ma, L. Xu, K.J. Chen, Physica E **41**, 923 (2009)
95. T. Zheng, Z. Li, Superlattices Microstruct. **37**, 227 (2005)
96. T. Gebel, L. Rebohle, J. Sun, W. Skorupa, Physica E **16**, 366 (2003)
97. O. Jambois, J. Carreras, A. Pérez-Rodríguez, B. Garrido, C. Bonafos, S. Schamm, G. Ben Assayag, Appl. Phys. Lett. **91**, 211105 (2007)
98. A.P. Baraban, Yu.V. Petrov, Semiconductors **42**, 1515 (2008)
99. L. Rebohle, J. von Borany, D. Borchert, H. Fröb, T. Gebel, M. Helm, W. Möller, W. Skorupa, Electrochem. Solid State Lett. **7**, G57 (2001)
100. L. Rebohle, T. Gebel, J. von Borany, W. Skorupa, M. Helm, D. Pacifici, G. Franzò, F. Priolo, Appl. Phys. B **74**, 53 (2002)

101. A.N. Nazarov, I.N. Osiyuk, J.M. Sun, R.A. Yankov, W. Skorupa, I.P. Tyagulskii, V.S. Lysenko, S. Prucnal, T. Gebel, L. Rebohle, Appl. Phys. B **87**, 129 (2007)
102. A. Kanjilal, L. Rebohle, M. Voelskow, W. Skorupa, M. Helm, Appl. Phys. Lett. **94**, 051903 (2009)
103. F.L. Bregolin, M. Behar, U.S. Sias, S. Reboh, J. Lehmann, L. Rebohle, W. Skorupa, J. Appl. Phys. **106**, 106103 (2009)
104. K.V. Shcheglov, C.M. Yang, K.J. Vahala, H.A. Atwater, Appl. Phys. Lett. **66**, 745 (1995)
105. J.Y. Zhang, X.L. Wu, X.M. Bao, Appl. Phys. Lett. **71**, 2505 (1997)
106. Y. Chen, G.Z. Ran, Y.K. Sun, Y.B. Wang, J.S. Fu, W.T. Chen, Y.Y. Gong, D.X. Wu, Z.C. Ma, W.H. Zong, G.G. Qin, Nucl. Instr. Meth. B **183**, 305 (2001)
107. H.S. Bae, T.G. Kim, C.N. Whang, S. Im, J.S. Yun, J.H. Song, J. Appl. Phys. **91**, 4078 (2002)
108. J.K. Shen, X.L. Wu, C. Tan, R.K. Yuan, X.M. Bao, Phys. Lett. A **300**, 307 (2002)
109. M.J. Lu, X.M. Wu, W.G. Yao, Mater. Sci. Eng. B **100**, 152 (2003)
110. J. Mayandia, T.G. Finstada, C. Henga, S. Fossa, H. Klette, J. Luminesc. **127**, 362 (2007)
111. M. Yang, T.P. Chen, L. Ding, J.I. Wong, Y. Liu, W.L. Zhang, S. Zhang, F. Zhu, W.P. Goh, Electrochem. Solid State Lett. **12**, H238 (2009)
112. G.G. Qin, C.L. Heng, G.F. Bai, K. Wu, C.Y. Li, Z.C. Ma, W.H. Zong, L.-p. You, Appl. Phys. Lett. **75**, 3629 (1999)
113. L. Tsybeskov, S.P. Duttagupta, K.D. Hirschman, P.M. Fauchet, K.L. Moore, D.G. Hall, Appl. Phys. Lett. **70**, 1790 (1997)
114. G.Z. Ran, Y. Chen, W.C. Qin, J.S. Fu, Z.C. Ma, W.H. Zong, H. Lu, J. Qin, G.G. Qin, J. Appl. Phys. **90**, 5835 (2001)
115. F. Iacona, D. Pacifici, A. Irrera, M. Miritello, G. Franzò, F. Priolo, D. Sanfilippo, G. Di Stefano, P.G. Fallica, Appl. Phys. Lett. **81**, 3242 (2002)
116. D. Pacifici, A. Irrera, G. Franzò, M. Miritello, F. Iacona, F. Priolo, Physica E **16**, 331 (2003)
117. A.J. Kenyon, Semicond. Sci. Technol. **20**, R65 (2005)
118. J.M. Sun, W. Skorupa, T. Dekorsy, M. Helm, A.M. Nazarov, Opt. Mater. **27**, 1050 (2005)
119. F. Iacona, A. Irrera, G. Franzó, D. Pacifici, I. Crudi, M. Miritello, C.D. Presti, F. Priolo, IEEE J. Sel. Top. Quant. **12**, 1596 (2006)
120. A. Kanjilal, L. Rebohle, W. Skorupa, M. Helm, Appl. Phys. Lett. **94**, 101916 (2009)
121. A. Irrera, M. Galli, M. Miritello, R. LoSavio, F. Iacona, G. Franzò, A. Canino, A.M. Piro, M. Belotti, D. Gerace, A. Politi, M. Liscidini, M. Patrini, D. Sanfilippo, P.G. Fallica, L.C. Andreani, F. Priolo, Physica E **41**, 891 (2009)
122. O. Jambois, Y. Berencen, K. Hijazi, M. Wojdak, A.J. Kenyon, F. Gourbilleau, R. Rizk, B. Garrido, J. Appl. Phys. **106**, 063526 (2009)
123. M. Yoshihara, A. Sekiya, T. Morita, K. Ishii, S. Shimoto, S. Sakai, Y. Ohki, J. Phys. D **30**, 1908 (1997)
124. S. Wang, H. Amekura, A. Eckau, R. Carius, Ch. Buchal, Nucl. Instrum. Method. B **148**, 481 (1999)
125. C. Buchal, Nucl. Instrum. Method. B **166–167**, 743 (2000)
126. J.M. Sun, W. Skorupa, T. Dekorsy, M. Helm, L. Rebohle, T. Gebel, J. Appl. Phys. **97**, 123513 (2005)
127. J.M. Sun, W. Skorupa, T. Dekorsy, M. Helm, L. Rebohle, T. Gebel, Appl. Phys. Lett. **85**, 3387 (2004)
128. S. Prucnal, J.M. Sun, W. Skorupa, M. Helm, Appl. Phys. Lett. **90**, 181121 (2007)
129. W. Skorupa, J.M. Sun, S. Prucnal, L. Rebohle, T. Gebel, A.N. Nazarov, I.N. Osiyuk, M. Helm, Solid State Phenom. **108–109**, 755 (2005)
130. J.M. Sun, S. Prucnal, W. Skorupa, M. Helm, L. Rebohle, T. Gebel, Appl. Phys. Lett. **89**, 091908 (2006)
131. L. Rebohle, J. Lehmann, A. Kanjilal, S. Prucnal, A. Nazarov, I. Tyagulskii, W. Skorupa, M. Helm, Nucl. Instrum. Methods Phys. Res. B **267**, 1324 (2009)
132. L. Rebohle, J. Lehmann, S. Prucnal, A. Nazarov, I. Tyagulskii, S. Tyagulskii, A. Kanjilal, M. Voelskow, D. Grambole, W. Skorupa, M. Helm, J. Appl. Phys. **106**, 123103 (2009)
133. M. Helm, J.M. Sun, J. Potfajova, T. Dekorsy, B. Schmidt, W. Skorupa, Microelectron. J. **36**, 957 (2005)

134. A.N. Nazarov, I.N. Osiyuk, I.P. Tyagulskii, V.S. Lysenko, S. Prucnal, J.M. Sun, W. Skorupa, R.A. Yankov, J. Luminesc. **121**, 213 (2006)
135. J.M. Sun, S. Prucnal, W. Skorupa, T. Dekorsy, A. Mücklich, M. Helm, L. Rebohle, T. Gebel, J. Appl. Phys. **99**, 103102 (2006)
136. S. Prucnal, J.M. Sun, A. Nazarov, I.P. Tyagulskii, I.N. Osiyuk, R. Fedaruk, W. Skorupa, Appl. Phys. B **88**, 241 (2007)
137. S. Prucnal, J.M. Sun, A. Mücklich, W. Skorupa, Electrochem. Solid State Lett. **10**, H50 (2007)
138. S. Prucnal, J.M. Sun, L. Rebohle, W. Skorupa, Appl. Phys. Lett. **91**, 181107 (2007)
139. S. Prucnal, J.M. Sun, H. Reuther, W. Skorupa, Ch. Buchal, Electrochem. Solid State Lett. **10**, J30 (2007)
140. S. Prucnal, J.M. Sun, H. Reuther, C. Buchal, J. uk, W. Skorupa, Vaccum **81**, 1296 (2007)
141. S. Wang, S. Coffa, R. Carius, C. Buchal, Mater. Sci. Eng. **B81**, 102 (2001)
142. J.M. Sun, L. Rebohle, S. Prucnal, M. Helm, W. Skorupa, Appl. Phys. Lett. **92**, 071103 (2008)
143. A. Irrera, M. Miritello, D. Pacifici, G. Franzò, F. Priolo, F. Iacona, D. Sanfilippo, G. Di Stefano, P.G. Fallica, Nucl. Instrum. Methods. Phys. Res. B **216**, 222 (2004)
144. S. Prucnal, L. Rebohle, A. Nazarov, I. Osiyuk, I. Tyagulskii, W. Skorupa, Appl. Phys. B **91**, 123 (2008)
145. M.E. Castagna, S. Coffa, M. Monaco, A. Muscara, L. Caristia, S. Lorenti, A. Messina, Mater. Sci. Eng. B **105**, 83 (2003)
146. A. Irrera, F. Iacona, G. Franzò, S. Boninelli, D. Pacifici, M. Miritello, C. Spinella, D. Sanfilippo, G. Di Stefano, P.G. Fallica, F. Priolo, Opt. Mater. **27**, 1031 (2005)
147. S. Prucnal, L. Rebohle, W. Skorupa, Appl. Phys. B **98**, 451 (2010)
148. L. Rebohle, T. Gebel, W. Skorupa, J.M. Sun, German Patent DE 10 2005 052 582 A1, 2005
149. B. El-Kareh, *Fundamentals of Semiconductor Processing Technology* (Kluwer, Boston, 1995)
150. D. Widmann, H. Mader, H. Friedrich, *Technology of Integrated Circuits* (Springer, Berlin, 2000)
151. J.F. Ziegler, J.P. Biersack, *The Stopping and Range of Ions in Matter* (Pergamon, New York, 2003)
152. R. Singh, J. Appl. Phys. **63**, R59 (1988)
153. W. Skorupa, T. Gebel, R.A. Yankov, S. Paul, W. Lerch, D.F. Downey, E.A. Arevalod, J. Electrochem. Soc. **152**, G436 (2005)
154. R. Ditchfield, E.G. Seebauer, J. Electrochem. Soc. **144**, 1842 (1997)
155. D.E. Mercer, A. Jain, S.W. Butler, Electrochem. Soc. Proc. **2001–9**, 247 (2001)
156. F.J. Grunthaner, P.J. Grunthaner, Mater. Sci. Rep. **1**, 65 (1986)
157. D.L. Griscom, J. Ceram. Soc. Jpn. **99**, 923 (1991)
158. W.L. Warren, E.H. Pointdexter, M. Offenberg, W. Müller-Warmuth, J. Electrochem. Soc. **139**, 872 (1992)
159. B. Garrido, J. Samitier, J.R. Morante, J. Montserrat, C. Domìnguez, Phys. Rev. B **49**, 14845 (1994)
160. N. Koshizaki, H. Umehara, T. Oyama, Thin Solid Films **325**, 130 (1998)
161. G. Pacchioni, L. Skuja, D.L. Griscom (eds.), *Defects in SiO_2 and Related Dielectrics: Science and Technology* (Kluwer, Dordrecht, 2000)
162. D.M. Fleedwood, S.T. Pantelides, R.D. Schrimpf, *Defects in Microelectronic Materials and Devices* (CRC, Boca Raton, 1991)
163. L. Skuja, in *Defects in SiO_2 and Related Dielectrics: Science and Technology*, ed. by G. Pacchioni, L. Skuja, D.L. Griscom (Kluwer, Dordrecht, 2000), p. 73
164. L.A. Christel, J.F. Gibbons, W.T. Sigmon, J. Appl. Phys. **52**, 7143 (1981)
165. E.P. EerNisse, C.B. Norris, J. Appl. Phys. **45**, 5196 (1974)
166. H. Hosono, K. Kawamura in *Defects in SiO_2 and related Dielectrics: Science and Technology*, ed. by G. Pacchioni, L. Skuja, D.L. Griscom (Kluwer, Dordrecht, 2000), p. 213
167. L.S. Liao, X.M. Bao, X.Q. Zheng, N.S. Li, N.B. Min, Appl. Phys. Lett. **68**, 850 (1996)
168. M. Fujinami, N.B. Chilton, Appl. Phys. Lett. **62**, 1131 (1993)
169. Z. Yin, F.W. Smith, Phys. Rev. B **43**, 4507 (1991)
170. D. Criado, M.I. Alayo, I. Pereyra, M.C.A. Fantini, Mater. Sci. Eng. B **112**, 123 (2004)

171. F.H.P.M. Habraken, A.E.T. Kuiper, Mater. Sci. Eng. **R12**, 123 (1994)
172. C.-M. Mo, L. Zhang, C. Xi, T. Wang, J. Appl. Phys. **73**, 5185 (1993)
173. H.L. Hao, L.K. Wu, W.Z. Shen, H.F.W. Dekkers, Appl. Phys. Lett. **91**, 201922 (2007)
174. A. Stoffel, A. Kovács, W. Kronast, B. Müller, J. Micromech. Microeng. **6**, 1 (1996)
175. C.M.M. Denisse, K.Z. Troost, F.H.P.M. Habraken, W.F. van der Weg, M. Hendriks, J. Appl. Phys. **60**, 2543 (1986)
176. A. del Prado, E. San Andres, I. Martil, G. Gonzalez-Diaz, D. Bravo, F.J. Lopez, M. Fernandez, F.L. Martinez, J. Appl. Phys. **94**, 1019 (2003)
177. F. Ay, A. Aydinli, Opt. Mater. **26**, 33 (2004)
178. W.D. Brown, M.A. Khaliq, Thin Solid Films **186**, 73 (1990)
179. R. Mroczynski, N. Kwietniewski, M. Cwil, P. Hoffmann, R.B. Beck, A. Jakubowski, Vacuum **82**, 1013 (2008)
180. A. Mrotzek, Bachelor thesis, West Saxon University of Applied Sciences of Zwickau, Germany, 2010
181. V.A. Gritsenko, H. Wong, J.B. Xu, R.M. Kwok, I.P. Petrenko, B.A. Zaitsev, Y.N. Morokov, Y.N. Novikov, J. Appl. Phys. **86**, 3234 (1999)
182. J.G. Zhu, C.W. White, J.D. Budai, S.P. Withrow, Y. Chen, J. Appl. Phys. **78**, 4386 (1995)
183. A. Markwitz, L. Rebohle, H. Hofmeister, W. Skorupa, Nucl. Instrum. Methods Phys. Res. B **147**, 361 (1999)
184. J. von Borany, K.H. Heinig, R. Grötzschel, M. Klimenkov, M. Strobel, K.H. Stegemann, H.J. Thees, Microelectron. Eng. **48**, 231 (1999)
185. A. Markwitz, R. Grötzschel, K.H. Heinig, L. Rebohle, W. Skorupa, Nucl. Instrum. Methods Phys. Res. B **152**, 319 (1999)
186. A. Polman, D.C. Jacobson, D.J. Eaglesham, R.C. Kistler, J.M. Poate, J. Appl. Phys. **70**, 3778 (1991)
187. S. Prucnal, J.M. Sun, L. Rebohle, W. Skorupa, Mater. Sci. Eng. B **146**, 241 (2008)
188. D.R. Lide (ed.), *Handbook of Chemistry and Physics* (CRC, Boca Raton, 2006)
189. M. Nagamori, J.-A. Boivin, A. Claveau, J. Noncryst. Sol. **189**, 270 (1995)
190. L. Rebohle, J. Lehmann, S. Prucnal, A. Kanjilal, A. Nazarov, I. Tyagulskii, W. Skorupa, M. Helm, Appl. Phys. Lett. **93**, 071908 (2008)
191. H. Ofuchi, Y. Imaizumi, H. Sugawara, H. Fujioka, M. Oshima, Y. Takeda, Nucl. Instrum. Methods Phys. Res. B **199**, 231 (2003)
192. J. Li, O.H.Y. Zalloum, T. Roschuk, C.L. Heng, J. Wojcik, P. Mascher, Appl. Phys. Lett. **94**, 011112 (2009)
193. R. Kögler, A. Mücklich, F. Eichhorn, M. Posselt, H. Reuther, W. Skorupa, J. Appl. Phys. **100**, 104314 (2006)
194. S.M. Sze, *Physics of Semiconductor Devices* (Wiley, New York, 1981)
195. E.H. Nicollian, J.R. Brews, *MOS (Metal Oxide Semiconductor) Physics and Technology* (Wiley, New York, 1982)
196. G. Barbottin, A. Vapaille (eds.), *Instabilities in Silicon Devices* (Elsevier, Amsterdam, 1986)
197. V.A. Gritsenko, N.D. Dikovskaja, K.P. Mogilnikov, Thin Solid Films **51**, 353 (1978)
198. C. Ance, F. de Chelle, J.P. Farraton, G. Leveque, P. Ordejón, F. Ynduráin, Appl. Phys. Lett. **60**, 1399 (1992)
199. K. Sugiyama, H. Ishii, Y. Ouchi, K. Seki, J. Appl. Phys. **87**, 295 (2000)
200. E. Centurioni, D. Iencinella, IEEE Electron. Dev. Lett. **24**, 177 (2003)
201. L. Gupta, A. Mansingh, P.K. Srivastava, Thin Solid Films **176**, 33 (1989)
202. M. Lenzlinger, E.H. Snow, J. Appl. Phys. **40**, 278 (1969)
203. M.A. Lampert, Peter Mark, *Current Injection in Solids* (Academic, New York, 1970)
204. D.J. DiMaria, M. Fischetti, in *Excess Electrons in Dielectric Media*, ed. by C. Ferradini, J.P. Jay-Gerin (CRC, Boca Raton, 1991), p. 315
205. H.-J. Fitting, J.-U. Friemann, Phys. Stat. Sol. A **69**, 349 (1982)
206. M.V. Fischetti, D.J. DiMaria, S.D. Brorson, T.N. Theis, J.R. Kirtley, Phys. Rev. B **31**, 8124 (1985)
207. S.D. Brorson, D.J. DiMaria, M.V. Fischetti, F.L. Pesavento, P.M. Solomon, D.W. Dong, J. Appl. Phys. **58**, 1302 (1985)

208. D. Arnold, E. Cartier, D.J. DiMaria, Phys. Rev. B **49**, 10278 (1994)
209. I.C. Chen, S. Holland, K.K. Young, C. Chang, C. Hu, Appl. Phys. Lett. **49**, 669 (1986)
210. D.J. DiMaria, D. Arnold, E. Cartier, Appl. Phys. Lett. **61**, 2329 (1992)
211. V.V. Afanas'ev, V.K. Adamchuk, Prog. Surf. Sci. **47**, 301 (1994)
212. A.N. Nazarov, T. Gebel, L. Rebohle, W. Skorupa, I.N. Osiyuk, V.S. Lysenko, J. Appl. Phys. **94**, 4440 (2003)
213. J.A. Lopez-Villanueva, J.A. Jimenez-Tejada, P. Cartujo, J. Bausells, J.E. Carceller, J. Appl. Phys. **70**, 3712 (1991)
214. A.N. Nazarov, S.I. Tiagulskyi, I.P. Tyagulskyy, V.S. Lysenko, L. Rebohle, J. Lehmann, S. Prucnal, M. Voelskow, W. Skorupa, J. Appl. Phys. **107**, 123112 (2010)
215. H.Y. Choi, H. Wong, V. Filip, B. Sen, C.W. Kok, M. Chan, M.C. Poon, Thin Solid Films **504**, 7 (2006)
216. J.L. Fay, J. Beluch, B. Despax, G. Sarrabayrouse, Jpn. J. Appl. Phys. **40**, 7 (2001)
217. V.J. Kapoor, D. Xu, R.S. Bailey, R.A. Turi, J. Electrochem. Soc. **139**, 915 (1992)
218. G. Lucovsky, Z. Jing, D.R. Lee, J. Vac. Sci. Technol. B **14**, 2832 (1996)
219. W.L. Warren, J. Kanicki, J. Robertson, E.H. Poindexter, P.J. McWhorter, J. Appl. Phys. **74**, 4034 (1993)
220. T. Hori, H. Iwasaki, Y. Yoshioka, M. Sato, Appl. Phys. Lett. **52**, 736 (1988)
221. H.J. Stein, J. Electrochem. Soc. **129**, 1786 (1982)
222. D.J. DiMaria, E. Cartier, D. Arnold, J. Appl. Phys. **73**, 3383 (1993)
223. A. Yankova, L. Do Thanh, P. Balk, Solid-St. Electron. **30**, 939 (1987)
224. M.V. Fischetti, J. Appl. Phys. **57**, 2860 (1985)
225. D.J. DiMaria, E. Cartier, D.A. Buchanan, J. Appl. Phys. **80**, 304 (1996)
226. C.T. Sah, J.Y.-C. Sun, J.J.-T. Tzou, J. Appl. Phys. **55**, 1525 (1984)
227. S.E. Thompson, T. Nishida, J. Appl. Phys. **72**, 4683 (1992)
228. D. Vuillaume, A. Bravaix, J. Appl. Phys. **73**, 2559 (1993)
229. V.A. Gritsenko, A.V. Shaposhnikov, Yu.N. Novikov, A.P. Baraban, H. Wong, G.M. Zhidomirov, M. Roger, Microelectron. Reliab. **43**, 665 (2003)
230. P. Balk, M. Aslam, D.R. Young, Solid State Electron. **27**, 709 (1984)
231. E.H. Nicollian, C.N. Berglund, P.F. Schmidt, J.M. Andrews, J. Appl. Phys. **42**, 5654 (1971)
232. D.J. DiMaria, M. Fischetti, E. Tierney, S.D. Brorson, Phys. Rev. Lett. **56**, 1284 (1986)
233. G.H. Dieke, *Spectra and Energy Levels of Rare Earth Ions in Crystals* (Interscience, New York, 1968)
234. G. Liu, B. Jacquier (eds.), *Spectroscopic Properties of Rare Earth in Optical Materials* (Tsinghua University Press and Springer, Berlin, 2005)
235. J. Rubio, J. Phys. Chem. Solids **52**, 101 (1991)
236. P. Dorenbos, J. Luminesc. **91**, 91 (2000)
237. P. Dorenbos, J. Luminesc. **91**, 155 (2000)
238. P. Dorenbos, J. Phys. Condens. Matter **15**, 575 (2003)
239. Simon Cotton, *Lanthanides and Actinides* (Macmillan, Basingstoke, 1991)
240. A.J. Steckl, J.C. Heikenfeld, D.-S. Lee, M.J. Garter, C.C. Baker, Y. Wang, R. Jones, IEEE Sel. Top. Quant. Electr. **8**, 749 (2002)
241. R. Tohmon, Y. Shimogaichi, H. Mizuno, Y. Ohki, K. Nagasawa, Y. Hama, Phys. Rev. Lett. **62**, 1388 (1989)
242. A.N. Trukhin, M. Goldberg, J. Jansons, H.-J. Fitting, I.A. Tale, J. Non-Cryst. Solids **223**, 114 (1998)
243. H. Nishikawa, E. Watanabe, D. Ito, Y. Sakurai, K. Nagasawa, Y. Ohki, J. Appl. Phys. **80**, 3513 (1996)
244. S.-T. Chou, J.-H. Tsai, B.-C. Sheu, J. Appl. Phys. **83**, 5394 (1998)
245. S. Munekuni, T. Yamanaka, Y. Shimogaichi, R. Tohmon, Y. Ohki, K. Nagasawa, Y. Hama, J. Appl. Phys. **68**, 1212 (1990)
246. J.C. Cheang-Wong, A. Oliver, J. Roiz, J.M. Hernández, L. Rodríguez-Fernández, J.G. Morales, A. Crespo-Sosa, Nucl. Instr. Methods. B **175–177**, 490 (2001)
247. G.-R. Lin, C.-J. Lin, C.-K. Lin, L.-J. Chou, Y.-L. Chueh, J. Appl. Phys. **97**, 094306 (2005)
248. S.V. Deshpande, E. Gulari, J. Appl. Phys. **77**, 6534 (1995)

249. K.J. Price, L.-R. Sharpe, L.E. Neil, E.A. Irene, J. Appl. Phys. **86**, 2638 (1999)
250. H. Kato, A. Masuzawa, H. Sato, T. Noma, K.S. Seol, M. Fujimaki, J. Appl. Phys. **90**, 2216 (2001)
251. Y. Xin, Y. Shi, H. Liu, Z.X. Huang, L. Pu, R. Zhang, Y.D. Zheng, Thin Sol. Films **516**, 1130 (2008)
252. R. Huang, K. Chen, P. Han, H. Dong, X. Wang, D. Chen, W. Li, J. Xu, Z. Ma, X. Huang, Appl. Phys. Lett. **90**, 093515 (2007)
253. N.M. Park, T.S Kim, S.J. Park, Appl. Phys. Lett. **78**, 2575 (2001)
254. K.S. Cho, N.M. Park, T.Y. Kim, K.H. Kim, G.Y. Sung, J.H. Shin, Appl. Phys. Lett. **86**, 071909 (2005)
255. K.S. Cho, N.M. Park, T.Y. Kim, K.H. Kim, J.H. Shin, G.Y. Sung, J.H. Shin, Appl. Phys. Lett. **88**, 209904 (2006)
256. L. Rebohle, W. Skorupa, Mater. Sci. Forum **590**, 117 (2008)
257. N. Miura, T. Sasaki, H. Matsumoto, Jpn. J. Appl. Phys. Part 2 **30**, L1815 (1991)
258. R.T. Wegh, H. Donker, A. Meijerink, R.J. Lamminmäki, J. Hölsä, Phys. Rev. B **56**, 13841 (1997)
259. C.A. Poynton, *Digital Video and HDTV: Algorithms and Interfaces* (Morgan Kaufmann, San Francisco, 2003), p. 205
260. A. Nazarov, J.M. Sun, W. Skorupa, R.A. Yankov, I.N. Osiyuk, I.P. Tjagulskii, V.S. Lysenko, T. Gebel, Appl. Phys. Lett. **86**, 151914 (2005)
261. E. Nakazawa, S. Shionoya, Phys. Rev. Lett. **25**, 1710 (1970)
262. G.S. Maciel, A. Biswas, R. Kapoor, P.N. Prasad, Appl. Phys. Lett. **76**, 1978 (2000)
263. S. Prucnal, W. Skorupa, L. Rebohle, M. Helm, German Patent DE 10 2007 019 209 A1, 2007
264. A. Polman, J. Appl. Phys. **82**, 1 (1997)
265. Y.Y. Choi, K.-S. Sohn, H.D. Park, S.Y. Choi, J. Mater. Res. **16**, 881 (2001)
266. H. Nakane, A. Noya, S. Kuriki, G. Matsumoto, Thin Solid Films, **59**, 291 (1979)
267. G. Adachi, N. Imanaka, Z.C. Kang (eds.), *Binary Rare Earth Oxides* (Kluwer, Dordrecht, 2004), p.117
268. G. Blasse, Physica Status Solidi. **55(b)**, K131 (1973)
269. C.D. Presti, A. Irrera, G. Franzò, I. Crupi, F. Prioloa, F. Iacona, G. Di Stefano, A. Piana, D. Sanfilippo, P.G. Fallica, Appl. Phys. Lett. **88**, 033501 (2006)
270. M. Fujita, Y. Tanaka, S. Noda, IEEE J. Sel. Top. Quantum Electron. **14**, 1090 (2008)
271. K. Bao, X.N. Kang, B. Zhang, T. Dai, C. Xiong, H. Ji, G.Y. Zhang, Y. Chen, IEEE Photonics Techn. Lett. **19**, 1840 (2007)
272. E.-J. Hong, K.-J. Byeon, H. Park, J. Hwang, H. Lee, K. Choi, H.-S. Kim, Solid State Electron. **53**, 1099 (2009)
273. M. Galli, A. Politi, M. Belotti, D. Gerace, M. Liscidini, M. Patrini, L.C. Andreani, M. Miritello, A. Irrera, F. Priolo, Y. Chen, Appl. Phys. Lett. **88**, 251114 (2006)
274. J. Potfajova, J.M Sun, B. Schmidt, T. Dekorsy, W. Skorupa, M. Helm, J. Luminesc. **121**, 290 (2006)
275. J. Vučković, M. Lončar, A. Scherer, IEEE J. Quantum Electron. **36**, 1131 (2000)
276. E. Fort, S. Grésillon, J. Phys. D **41**, 013001 (2008)
277. K. Okamoto, Y. Kawakami, IEEE J. Sel. Top. Quantum Electron. **15**, 1199 (2009)
278. L. Rebohle, J. Lehmann, S. Prucnal, J.M. Sun, M. Helm, W. Skorupa, Appl. Phys. B **98**, 439 (2010)
279. A.J. Kenyon, P.F. Trwoga, M. Federighi, C.W. Pitt, J. Phys. Condens. Matter **6**, L319 (1994)
280. G. Franzó, V. Vinciguerra, F. Priolo, Appl. Phys. **A 69**, 3 (1999)
281. G. Franzó, V. Vinciguerra, F. Priolo, Philos. Mag. B **80**, 719 (2007)
282. G. Franzó, E. Pecora, F. Priolo, Appl. Phys. Lett. **90**, 183102 (2007)
283. P.G. Kik, A. Polman, J. Appl. Phys. **88**, 1992 (2000)
284. D. Pacifici, G. Franzó, F. Priolo, F. Iacona, L. Dal Negro, Phys. Rev. B **67**, 245301 (2003)
285. D. Pacifici, L. Lanzanò, G. Franzò, F. Priolo, F. Iacona, Phys. Rev. B **72**, 045349 (2005)
286. M. Wojdak, M. Klik, M. Forcales, O.B. Gusev, T. Gregorkiewicz, D. Pacifici, G. Franzó, F. Priolo, F. Iacona, Phys. Rev. B **69**, 233315 (2004)

287. F. Gourbilleau, M. Levalois, C. Dufour, J. Vicens, R. Rizk, J. Appl. Phys. **95**, 3717 (2004)
288. S. Minissale, T. Gregorkiewicz, M. Forcales, R.G. Elliman, Appl. Phys. Lett. **89**, 171908 (2006)
289. D. Kovalev, J. Diener, H. Heckler, G. Polisski, N. Künzner, F. Koch, Phys. Rev. B **61**, 4485 (2000)
290. B. Garrido, C. García, P. Pellegrino, D. Navarro-Urrios, N. Daldosso, L. Pavesi, F. Gourbilleau, R. Rizk, Appl. Phys. Lett. **89**, 163103 (2006)
291. C.E. Chryssou, A.J. Kenyon, T.S. Iwayama, C.W. Pitt, D.E. Hole, Appl. Phys. Lett. **75**, 2011 (1999)
292. K. Imakita, M. Fujii, Y. Yamaguchi, S. Hayashi, Phys. Rev. B **71**, 115440 (2005)
293. I. Izeddin, A.S. Moskalenko, I.N. Yassievich, M. Fujii, T. Gregorkiewicz, Phys. Rev. Lett. **97**, 207401 (2006)
294. I. Izeddin, Dissertation, University of Amsterdam, The Netherlands, 2008
295. M. Fujii, K. Imakita, K. Watanabe, S. Hayashi, J. Appl. Phys. **95**, 272 (2004)
296. O. Savchyn, F.R. Ruhge, P.G. Kik, R.M. Todi, K.R. Coffey, H. Nukala, H. Heinrich, Phys. Rev. B **76**, 195419 (2007)
297. A. Kanjilal, L. Rebohle, M. Voelskow, W. Skorupa, M. Helm, J. Appl. Phys. **104**, 103522 (2008)
298. S. Núñez-Sánchez, R. Serna, J.G. López, A.K. Petford-Long, M. Tanase, B. Kabius, J. Appl. Phys. **105**, 013118 (2009)
299. L. Dal Negro, J.H. Yi, M. Hiltunen, J. Michel, C. Kimerling, S. Hamel, A.J. Williamson, G. Galli, T.-W.F. Chang, V. Sukhovatkin, E.H. Sargent, J. Exp. Nanosci. **1**, 29 (2006)
300. R. Li, S. Yerci, L. Dal Negro, Appl. Phys. Lett. **95**, 041111 (2009)
301. C. Delerue, G. Allan, M. Lannoo, Phys. Rev. B **48**, 11024 (1993)
302. F. Iacona, G. Franzó, C. Spinella, J. Appl. Phys. **87**, 1301 (2000)
303. F. Priolo, C.D. Presti, G. Franzò, A. Irrera, I. Crupi, F. Iacona, G. Di Stefano, A. Piana, D. Sanfilippo, P.G. Fallica, Phys. Rev. B **73**, 113302 (2006)
304. K. Sun, W.J. Xu, B. Zhang, L.P. You, G.Z. Ran, G.G. Qin, Nanotechnology **19**, 105708 (2008)
305. C. Bulutay, Phys. Rev. B **76**, 205321 (2007)
306. J.H. Chen, D. Pang, H.M. Cheong, P. Wickboldt, W. Paul, Appl. Phys. Lett. **67**, 2182 (1995)
307. C.L. Heng, T.G. Finstad, P. Storås, Y.J. Li, A.E. Gunnæs, Appl. Phys. Lett. **85**, 4475 (2004)
308. C.L. Heng, E. Chelomentsev, Z.L. Peng, P. Mascher, P.J. Simpson, J. Appl. Phys. **105**, 014312 (2009)
309. J.S. Jensen, T.P.L. Pedersen, J. Chevallier, B.B. Nielsen, A.N. Larsen, Nanotechnology **17**, 2621 (2006)
310. A. Kanjilal, S. Tsushima, C. Götz, L. Rebohle, M. Voelskow, W. Skorupa, M. Helm, J. Appl. Phys. **106**, 063112 (2009)
311. A. Kanjilal, L. Rebohle, N.K. Baddela, S. Zhou, M. Voelskow, W. Skorupa, M. Helm, Phys. Rev. B **79**, 161302(R) (2009)
312. A. Kanjilal, L. Rebohle, M. Voelskow, W. Skorupa, M. Helm, J. Appl. Phys. **106**, 026104 (2009)
313. X.L. Wu, T. Gao, X.M. Bao, F. Yan, S.S. Jiang, D. Feng, J. Appl. Phys. **82**, 2704 (1997)
314. H. Ninomiya, N. Itoh, S. Rath, S. Nozaki, H. Morisaki, J. Vac. Sci. Technol. B **17**, 1903 (1999)
315. P. Pellegrino, B. Garrido, J. Arbiol, C. Garcia, Y. Labour, J.R. Morante, Appl. Phys. Lett. **88**, 121915 (2006)
316. A.K. Arora, M. Rajalakshmi, T.R. Ravindran, V. Sivasubramanian, J. Raman Spectrosc. **38**, 604 (2007)
317. J.I. Langford, A.J.C. Wilson, J. Appl. Crystallogr. **11**, 102 (1978)
318. E. Morosan, J.A. Fleitman, Q. Huang, J.W. Lynn, Y. Chen, X. Ke, M.L. Dahlberg, P. Schiffer, C.R. Craley, R.J. Cava, Phys. Rev. B **77**, 224423 (2008)
319. M.C. Ridgway, G. De, M. Azevedo, R.G. Elliman, C.J. Glover, D.J. Llewellyn, R. Miller, W. Wesch, G.J. Foran, J. Hansen, A. Nylandsted-Larsen, Phys. Rev. B **71**, 094107 (2005)
320. Q. Xu, I.D. Sharp, C.W. Yuan, D.O. Yi, C.Y. Liao, A.M. Glaeser, A.M. Minor, J.W. Beeman, M.C. Ridgway, P. Kluth, J.W. Ager III, D.C. Chrzan, E.E. Haller, Phys. Rev. Lett. **97**, 155701 (2006)

321. H. Suzuki, T.A. Tombrello, C.L. Melcher, J.S. Schweitzer, Nucl. Instrum. Methods Phys. Res. A **320**, 263 (1992)
322. H. Guo, Y. Li, D. Wang, W. Zhang, M. Yin, L. Lou, S. Xia, J. Alloys Compounds **376**, 23 (2004)
323. H. Xu, Z. Jiang, Phys. Rev. B **66**, 035103 (2002)
324. F. Auzel, Y. Chen, D. Meichenin, J. Luminesc. **60–61**, 692 (1994)
325. H. Hosono, M. Mizuguchi, L. Skuja, T. Ogawa, Opt. Lett. **24**, 1549 (1999)
326. X.W. Wang, T.P. Ma, Appl. Phys. Lett. **60**, 2634 (1992)
327. P.J. Wright, N. Kasai, S. Inoue, K. Saraswat, IEEE Electron. Device Lett. **10**, 347 (1989)
328. K.P. MacWilliams, L.F. Halle, T.C. Zietlow, IEEE Electron. Dev. Lett. **11**, 3 (1990)
329. D.S. Kim, J. Lee, K. Char, Appl. Phys. Lett. **87**, 042107 (2005)
330. Y. Nishioka, K. Ohyu, Y. Ohji, N. Natuaki, K. Mukai, T.P. Ma, IEEE Electron. Dev. Lett. **10**, 141 (1989)
331. S. Shen, A. Iha, Opt. Mater. **25**, 321 (2004)
332. K. Takahashi, J. Miyahara, Y. Shibahara, J. Electrochem. Soc. **132**, 1492 (1985)
333. H. von Seggern, T. Voigt, W. Knupfer, G. Lange, J. Appl. Phys. **64**, 1405 (1988)
334. S. Prucnal, Dissertation, Maria Curie-Skłodowska University Lublin, Poland, 2008
335. J.H. Stathis, IEEE Trans. Dev. Mater. Reliab. **1**, 43 (2001)
336. P.M. Lenahan, P.V. Dressenhofer, J. Appl. Phys. **55**, 3498 (1984)
337. J.S. Suehle, in *Defects in Microelectronic Materials and Devices*, ed. by D.M. Fleetwood, S.T. Pantelides, R.D. Schrimpf (CRC, Boca Raton, 2009), p. 437
338. J.W. McPherson, R.B. Khamankar, Semicond. Sci. Technol. **15**, 462 (2000)
339. R. Gale, F.J. Feigl, C.W. Magee, D.R. Young, J. Appl. Phys. **54**, 6938 (1983)
340. G.V. Gadiyak, J. Appl. Phys. **82**, 5573 (1997)
341. J. Suñé, E.Y. Wu, Phys. Rev. Lett. **92**, 87601 (2004)
342. B. Kaczer, R. Degraeve, N. Pangon, G. Groeseneken, IEEE Trans. Electron Dev. **47**, 1514 (2000)
343. J. Suñé, I. Placencia, N. Barniol, E. Farrés, F. Martín, X. Aymerich, Thin Solid Films **185**, 347 (1990)
344. A. Aassime, Thin Solid Films **385**, 252 (2001)
345. E. Hasegawa, A. Ishitani, K. Akimoto, M. Tsukiji, N. Ohta, J. Electrochem. Soc. **142**, 273 (1995)
346. E. Cartier, J.H. Stathis, D.A. Buchanan, Appl. Phys. Lett. **63**, 1510 (1993)
347. R.E. Stahlbush, E. Cartier, IEEE Trans. Nucl. Sci. **41**, 1844 (1994)
348. Y. Nissan-Cohen, T. Gorczyca, IEEE Electron Dev. Lett. **9**, 287 (1988)
349. J. Suñé, E.Y. Wu, in *Defects in Microelectronic Materials and Devices*, ed. by D.M. Fleetwood, S.T. Pantelides, R.D. Schrimpf (CRC, Boca Raton, 2009), p. 465
350. A. Nazarov, W. Skorupa, I.N. Osiyuk, I.P. Tjagulskii, V.S. Lysenko, R.A. Yankov, T. Gebel, J. Electrochem. Soc. **152**, F20 (2005)
351. S. Prucnal, L. Rebohle, W. Skorupa, Appl. Phys. B **94**, 289 (2009)
352. S. Tyagulskiy, I. Tyagulskiy, A. Nazarov, V. Lysenko, L. Rebohle, J. Lehmann, W. Skorupa, Microelectron. Eng. **86**, 1954 (2009)
353. A.N. Trukhin, in *Defects in SiO_2 and Related Dielectrics: Science and Technology*, ed. by G. Pacchioni, L. Skuja, D.L. Griscom (Kluwer, Dordrecht, 2000), p. 235
354. J. Bautista, M. Morse, J. Swift (primary authors), 2005 Communications Technology Roadmap, MIT Microphotonics Center Industry Consortium, http://www.signallake.com/innovation/MicrophotonicsComm Roadmap2005.pdf
355. C.J.M. Arts, M.J. van Baak, C.J. Elliot, S.A. Hewitt, J. Cooper, K. van de Velde-Fasea, R.F. Witkampa, Analyst **123**, 2579 (1998)
356. S. Rodriguez, S. Reder, M. Lopez de Alda, G. Gauglitz, D. Barceló, Biosens. Bioelectron. **19**, 633 (2004)
357. J. Tschmelak et al., Int. J. Environ. Anal. Chem. **85**, 837 (2005)
358. R.F. Taylor, J.S. Schultz (eds.), *Handbook of Chemical and Biological Sensors* (IOP, Bristol, 1996)

359. F.S. Ligler, C.A. Rowe-Taitt (eds.), *Optical Biosensors: Present and Future* (Elsevier, Amsterdam, 2002)
360. P.N. Prasad (ed.) *Introduction to Biophotonics* (Wiley, Hoboken, 2003)
361. L. Rebohle, T. Gebel, R.A. Yankov, T. Trautmann, W. Skorupa, J. Sun, G. Gauglitz, R. Frank, Opt. Mater. **27**, 1055 (2005)
362. Y. Cai, R. Shinar, Z. Zhou, J. Shinar, Sens. Actuators B **134**, 727 (2008)
363. K. Awazu, C. Rockstuhl, M. Fujimaki, N. Fukuda, J. Tominaga, T. Komatsubara, T. Ikeda, Y. Ohki, Opt. Exp. **15**, 2592 (2007)
364. K.H.A. Lau, L.-S. Tan, K. Tamada, M.S. Sander, W. Knoll, J. Phys. Chem. B **108**, 10812 (2004)
365. Z.-M. Qi, I. Honma, H. Zhou, Appl. Phys. Lett. **90**, 011102 (2007)
366. Y. Jiao, S.M. Weiss, Biosens. Bioelectron. **25**, 1535 (2010)
367. E. Kretschmann, Z. Physik **241**, 313 (1971)
368. G. Rong, J.D. Ryckman, R.L. Mernaugh, S.M. Weiss, Appl. Phys. Lett. **93**, 161109 (2008)
369. H. Ouyang, C.C. Striemer, P.M. Fauchet, Appl. Phys. Lett. **88**, 163108 (2006)
370. T.S. Hug, J.E. Prenosil, P. Maier, M. Morbidelli, Biotechnol. Bioeng. **80**, 213 (2002)
371. C. Pacholski, M. Sartor, M.J. Sailor, F. Cunin, G.M. Miskelly, J. Am. Chem. Soc. **127**, 11636 (2005)
372. V.S.-Y. Lin, K. Motesharei, K.-P.S. Dancil, M.J. Sailor, M.R. Ghadiri, Science **278**, 840 (1997)
373. L. De Stefano, L. Moretti, A. Lamberti, O. Longo, M. Rocchia, A.M. Rossi, P. Arcari, I. Rendina, IEEE Trans. Nanotechnol. **3**, 49 (2004)
374. K.-P.S. Dancil, D.P. Greiner, M.J. Sailor, J. Am. Chem. Soc. **121**, 7925 (1999)
375. A. Tinsley-Bown, R.G. Smith, S. Hayward, M.H. Anderson, L. Koker, A. Green, R. Torrens, A.-S. Wilkinson, E.A. Perkins, D.J. Squirrell, S. Nicklin, A. Hutchinson, A.J. Simons, T.I. Cox, Phys. Stat. Sol. A **202**, 1347 (2005)
376. A. Janshoff, K.-P.S. Dancil, C. Steinem, D.P. Greiner, V.S.-Y. Lin, C. Gurtner, K. Motesharei, M.J. Sailor, M.R. Ghadiri, J. Am. Chem. Soc. **120**, 12108 (1998)
377. M. Holgado, R. Casquel, C. Molpeceres, M. Morales, J.L. Ocaña, Sensor Lett. **6**, 564 (2008)
378. J. Gao, T. Gao, M.J. Sailor, Appl. Phys. Lett. **77**, 901 (2000)
379. S. Zangooie, R. Bjorklund, H. Arwin, Thin Solid films **313**, 825 (1998)
380. P.A. Snow, E.K. Squire, P. St, J. Russell, L.T. Canham, J. Appl. Phys. **86**, 1781 (1999)
381. L.A. DeLouise, P.M. Kou, B.L. Miller, Anal. Chem. **77**, 3222 (2005)
382. F. Cunin, T.A. Schmedake, J.R. Link, Y.Y. Li, J. Koh, S.N. Bhatia, M.J. Sailor, Nat. Mater. **1**, 39 (2002)
383. K.A. Kilian, T. Böcking, K. Gaus, M. Gal, J.J. Gooding, ACS Nano **1**, 355 (2007)
384. M.S. Salem, M.J. Sailor, K. Fukami, T. Sakka, Y.H. Ogata, J. Appl. Phys. **103**, 083516 (2008)
385. S. Chan, Y. Li, L.J. Rothberg, B.L. Miller, P.M. Fauchet, Mater. Sci. Eng. C **15**, 277 (2001)
386. L. Tay, N.L. Rowell, D.J. Lockwood, R. Boukherroub, J. Vac. Sci. Technol. A **24**, 747 (2006)
387. V.M. Starodub, L.L. Fedorenko, A.P. Sisetskiy, N.F. Starodub, Sens. Actuators B **58**, 409 (1999)
388. F.P. Mathew, E.C. Alocilja, Biosens. Bioelectron. **20**, 1656 (2005)
389. K.M. Mayer, S. Lee, H. Liao, B.C. Rostro, A. Fuentes, P.T. Scully, C.L. Nehl, J.H. Hafner, ACS Nano **2**, 687 (2008)
390. H.M. Hiep, H. Yoshikawa, M. Saito, E. Tamiya, ACS Nano **3**, 446 (2009)
391. J. Homola, Anal. Bioanal. Chem. **377**, 528 (2003)
392. M. Lee, P.M. Fauchet, Opt. Exp. **15**, 4530 (2007)
393. S. Zlatanovic, L.W. Mirkarimi, M.M. Sigalas, M.A. Bynum, E. Chow, K.M. Robotti, G.W. Burr, S. Esener, A. Grot, Sens. Actuators B **141**, 13 (2009)
394. E. Chow, A. Grot, L.W. Mirkarimi, M. Sigalas, G. Girolami, Opt. Lett. **29**, 1093 (2004)
395. C.A. Barrios, M.J. Bañuls, V. González-Pedro, K.B. Gylfason, B. Sánchez, A. Griol, A. Maquieira, H. Sohlström, M. Holgado, R. Casquel, Opt. Lett. **33**, 708 (2008)
396. S.M. Weiss, G. Rong, J.L. Lawrie, Physica E **41** (2009) 1071
397. M.P. Stewart, J.M. Buriak, Adv. Mater. **12**, 859 (2000)

398. J.J. Shi, Y.F. Zhu, X.R. Zhang, W.R.G. Baeyens, A.M. Garcia-Campana, Trends Anal. Chem. **23**, 351 (2004)
399. A. Jane, R. Dronov, A. Hodges, N.H. Voelcker, Trends Biotechnol. **27**, 230 (2009)
400. C. Cherkouk, L. Rebohle, W. Skorupa, T. Strache, H. Reuther, M. Helm, J. Colloid Interf. Sci. **337**, 375 (2009)
401. C. Cherkouk, L. Rebohle, W. Skorupa, German Patent No. DE 10 2008 024 526 A1, 2008
402. L. Rebohle, C. Cherkouk, S. Prucnal, M. Helm, W. Skorupa, Vacuum **83**, S24 (2009)
403. C. Cherkouk, Dissertation, Technical University of Dresden, Germany, 2010
404. Molecular Probes, Inc
405. U. Hilleringmann, K. Goser, IEEE Trans. Electron. Dev. **42**, 841 (1995)
406. G. Maiello, M. Balucani, V. Bondarenko, G. De Cesare, S. La Monica, G. Masini, A. Ferrari, Solid State Phen. **54**, 45 (1997)
407. S.J. Koester, C.L. Schow, L. Schares, G. Dehlinger, J.D. Schaub, F.E. Doany, R.A. John, J. Lightwave Techn. **25**, 46 (2007)
408. K.-W. Ang, T.-Y. Liow, M.-B. Yu, Q. Fang, J. Song, G.-Q. Lo, D.-L. Kwong, IEEE J. Sel. Top. Quantum Electron. **16**, 106 (2010)
409. R. Yan, S.P. Mestas, G. Yuan, R. Safaisini, D.S. Dandy, K.L. Lear, Lab Chip **9**, 2163 (2009)
410. D. Navarro-Urrios, N. Daldosso, C. García, P. Pellegrino, B. Garrido, F. Gourbilleau, R. Rizk, L. Pavesi, Jpn. J. Appl. Phys. **46**, 6626 (2007)
411. N.M. Jokerst, M.A. Brooke, S.-Y. Cho, S. Wilkinson, M. Vrazel, S. Fike, J. Tabler, Y.J. Joo, S.-W. Seo, D.S. Wills, A. Brown, IEEE J. Sel. Top. Quantum Electron. **9**, 350 (2003)
412. M.C. Gather, F. Ventsch, K. Meerholz, Adv. Mater. **20**, 1966 (2008)
413. M. Ramuz, L. Bürgi, R. Stanley, C. Winnewisser, J. Appl. Phys. **105**, 084508 (2009)
414. M. Ramuz, D. Leuenberger, R. Pfeiffer, L. Bürgi, C. Winnewisser, Eur. Phys. J. Appl. Phys. **46**, 12510 (2009)
415. A. Yamada, M. Sakuraba, J. Murota, Mat. Sci. Semicond. Process. **8**, 435 (2005)
416. A. Yamada, M. Sakuraba, J. Murota, Thin Solid Films **508**, 399 (2006)
417. L. Rebohle, W. Skorupa, German Patent No. DE 10 2008 037 225 A1, 2008
418. Y.S. Chu, W.-H. Hsu, C.-W. Lin, W.-S. Wang, Microwave Opt. Techn. Lett. **48**, 955 (2006)
419. L.M.Lechuga et al., Sens. Actuators B **118**, 2 (2006)

Index

MIX
Papier aus verantwortungsvollen Quellen
Paper from responsible sources
FSC® C105338

If you have any concerns about our products,
you can contact us on
ProductSafety@springernature.com

In case Publisher is established outside the EU,
the EU authorized representative is:
Springer Nature Customer Service Center GmbH
Europaplatz 3, 69115 Heidelberg, Germany

Printed by Libri Plureos GmbH
in Hamburg, Germany